Introducing Volcanology

SECOND EDITION

Other Titles in this Series:

Introducing Volcanology

A Guide to Hot Rocks

SECOND EDITION

Dougal Jerram

DUNEDIN

EDINBURGH ◆ LONDON

This book is dedicated to Jo and Izabel
(Izzy – the Great Cornsteiny aka Mini Dr Volcano)
Volcano . . . BOOM!

Published by
Dunedin Academic Press Ltd
Hudson House
8 Albany Street
Edinburgh EH1 3QB
Scotland

www.dunedinacademicpress.co.uk

ISBNs
9781780460901 (Paperback)
9781780466552 (ePub)
9781780466569 (Amazon Kindle)
9781780466576 (PDF)

British Library Cataloguing in Publication data
A catalogue record for this book is available from the British Library

Typeset by Biblichor Ltd, Edinburgh
Printed and bound in Poland by Hussar Books

Contents

List of Illustrations and Tables

Illustration Credits and Acknowledgements

Unless otherwise indicated the photographs, tables and other figures used are those of the author.

The following illustrations are reproduced by permission, courtesy of:
Figure 1.1d: Josef Schlach; Figures 2.1, 3.4, 4.2, 4.5, 5.3, 7.11b, 8.3, 10.12, 10.13: USGS; Figures 1.2 and 6.1: Rick Hoblitt, USGS; Figures 1.8 and 4.7: Myers, B. and Driedger, C. USGS General Information Products 63 and 64; Figure 4.8a: Ed Klimasauskas *et al.*, USGS Fact Sheet 092-02; Figure 4.8b-c: David W. Ramsey *et al.*, USGS Geologic Investigations Series I-2790; Figure 6.5: Peter W. Lipman, USGS (image 113); Figure 8.5: Dave Harlow, USGS; Figure 10.8b: Lyn Topinka, USGS. Figures 1.4, 4.1, 5.1 and 10.10: Jón Viðar Sigurðsson; Figure 2.4: Lori Dickson; Figure 2.6: a) Jon Davidson, b) Adrian Pingstone; Figure 2.5: J.A. Karson & R.J. Wysocki, unpublished data, Syracuse University Lava Project (lavaproject.syr.edu). Figures 3.5, 3.6, 4.6, 5.10, 6.1, 8.6, 9.4, 10.3, 10.9: NASA/JPL; Figure 3.5: Malin Space Science Systems; Figure 4.3: Ivan d'Hostingue (aka Sémhur on Wikimedia Commons, © CC-BY-SA 3.0); Figure 5.8c: Morgan Jones; Figures 5.9 and 6.10b-d: Joseph Resing, Marine Geoscience Data System; Figure 5.11b: Sverre Planke; Figure 6.10a: Howell Williams, U.S. National Oceanic and Atmospheric Administration; Figure 6.11b: NASA/JPL/ University of Arizona; Figures 7.4, 7.5 and 7.14a: Olivier Galland; Figure 7.6: Mark Cooper, Geological Survey of Northern Ireland, Tellus project, www.bgs.ac.uk/gsni/tellus/index.html, *see* Cooper *et al.* (2012) *Journal of the Geological Society*, London. Vol. 169; Figure 7.7, Sergio Rocchi; Figure 7.11a: Schmidt Ocean Institute, ROV ROPOS; Figure 7.12a: I. Bothar, Wikimedia Commons © CC-BY-SA 3.0; Figure 7.14b: Sverre Planke; Figure 8.1: Henrik Svensen; Figure 8.4: Vincent Courtillot (*see also* V. Courtillot, F. Fluteau et A.L. Chenet, *Des éruptions volcaniques ravageuses*, Dossier Pour la Science N°51, avril–juin 2006); Figure 8.7: Galapagos map courtesy of Denis Geist; Figure 9.2: Jurgen Neuberg, data courtesy Montserrat Volcano Observatory; Figure 9.4: Julie Roberge; Figure 9.7b & c: William Hutchison and Nick Varley, Centre of Exchange and Research in Volcanology, Universidad de Colima, –www.ucol.mx/ciiv; Figure 9.9b: 3D model from John Howell; Figure 10.4: Martin C. Doege (based on NASA SRTM3 data); Figure 10.5: Media from the Discovery Channel's *Pompeii: The Last Day*, courtesy of Crew Creative Ltd; Figure 10.8a: Jim Nieland, US Forest Service; Figure 10.11a: Landsat 8 (NASA); Figure 10.11b: Bæring Steinþórsson. Figure 10.14: Indonesian National Armed Forces; Figure 10.16b: Sara Sorrentino; Figure 10.16c: John Howell.

The following illustrations have been adapted from published sources:
Figure 3.1: Graham Park (2010) *Introducing Geology*, Dunedin Academic Press. Figure 4.10: Image courtesy of Mike Coffin, from: Coffin, M.F. *et. al.* (2006) *Oceanography* 19(4), 150–160. 4.9a: 3D schematic image courtesy of Freysteinn Sigmundsson. Figure 6.3: Lockwood, J.P. and Hazlett, R.W. (2010) *Volcanoes: Global Perspective*s, Wiley-Blackwell. Figure 6.7: Sparks *et al.* (1973) *Geology*; 1(3), 115–118. Figure 7.2c: Image courtesy of Nick Schofield, from: Goulty, N. and Schofield, N. (2008) *Journal of Structural Geology*, 30, 812–817. Figure 7.8: Emeleus, C.H. & Bell, B.R. (2005) *British Regional Geology: the Palaeogene Volcanic Districts of Scotland.* British Geological Survey. Figure 7.9: Jerram, D. and Goodenough, K. (2008) Golden Rum, bicentenary field report, *Geoscientist* Volume 18, No. 3. https://www.geolsoc.org. uk/Geoscientist/Archive/March-2008/Golden-Rum. Figure 7.10: Ryan, M.P. (1988) *Journal of Geophysical Research*, 93, 4213–4248. Figure 8.2: White, R.V. and Saunders, A.D. (2005) *Lithos*, 79, 299–316. Figures 9.5, 9.6: Images courtesy of Freysteinn Sigmundsson from Sigmundsson *et al.* (2010) *Nature*, 468, 426–431. Figure 9.8: Jerram, D. and Smith, S. (2010) Earth's Hottest Place. *Geoscientist*, Volume 20, No. 2, 12–13 https://www.geolsoc.org.uk/Geoscientist/Archive/February-2010/Earths-hottest-place.

In addition, Catherine Nelson, Richard Brown, Olivier Galland, John Howell, Sverre Planke and Graham Andrews are thanked for their valuable input. Brian Upton and an anonymous reviewer are also thanked for their constructive and helpful comments.

Preface

Volcanoes have the power to rock our world, from the spectacular and the beautiful to the violent and the deadly. *Introducing Volcanology: A Guide to Hot Rocks* is a detailed but accessible introduction to volcanoes and their plumbing systems. Aimed at those with an inquisitive interest in volcanoes as well as the more advanced reader, the ten chapters document different aspects of volcanology. All are illustrated with a wide array of photographs and diagrams to accompany the text, and an A–Z of volcanology is included as a glossary. Since the first edition of this book was published in 2011, a lot has happened in the world of volcanoes and in the world of DougalEARTH. I have been lucky to continue travelling globally to see many of the world's modern and ancient volcanic rocks (the 'HOT Rocks' that form the basis of this guide). Since appearing as 'Dr Volcano' on BBC television, I have continued to work with various media projects including CBBC series *Fierce Earth*, a trip down the Grand Canyon in a wooden boat (BBC/Discovery) and a journey to the very edge of China (*Discovery Asia*), always looking out for the next Earth Science adventure.

I now have a number of books on volcanic rocks and more general aspects of Earth and popular science (including: *The Field Description of Igneous Rocks* – Wiley; *Volcanoes of Europe* – Dunedin; *Travel Guide to the Centre of the Earth* – Pilazzo; *Victor the Volcano* – Rudling House; *Dig to the Centre of the Earth* – Carlton). Of these books *Introducing Volcanology: A Guide to Hot Rocks*, remains my first solo venture in book writing and one I am very proud of. The first edition has topped the Amazon selling lists in 'Volcanoes, Earthquakes and Tectonics' on a few occasions, proving popular with interested volcanophiles and budding Earth scientists, and I hope the second edition is also received favourably. Given its previous popularity, I have chosen not to change/deviate too much in this edition, but moreover to update and enhance what we had before. More recent eruptions, as well as a trip I took to the Galápagos, have all inspired me to add some of the new materials, which along with the detailed colour images and graphics from before, will guide you through the world of hot rocks. Thanks also go to Anthony Kinahan and Anne Morton for help and motivation in putting the new version of *Introducing Volcanology* together, and to the varied people and institutions that have made pictures and figures available for use (*see* acknowledgements/credits section and in figure captions). I would also like to acknowledge the late David McLeod, who helped with the design and typesetting of the first edition of this book, as well as the second edition of *Volcanoes of Europe* also published by Dunedin. I hope you find *Introducing Volcanology: A Guide to Hot Rocks* a very informative text and an even more valuable source of images and graphics to help you develop your understanding of volcanoes, and inspire you to delve more deeply into Earth Science and the wonderful knowledge it can bring.

A final closing message that I would like to add to this preface of the second edition: two close colleagues and mentors to my career as an Earth scientist/volcanologist sadly passed away between the editions of this book – Jon Davidson (2016) and Henry Emeleus (2017). They both contributed so much to my understanding of igneous rocks and volcanoes, and I owe a great deal to them not only for their knowledge sharing, but also for the fond memories of times shared musing over rock samples and thin-sections, out in the field on geo-expeditions, or sharing fantastic yarns over a beer – cheers!!

Prof. Dougal A. Jerram, aka 'Dr Volcano', August 2020

Figure 1.1 Volcanoes in our lives. **a)** Painting of a child's idea of a volcano (Fleur Meston 11 yrs). **b)** 3D volcano model for a school competition (Tyron Wilson 12 yrs). **c)** Spot the eruption! Author on location in Stromboli for a filming project (Stromboli erupting in background). **d)** Volcanoes featured on stamps. (Stamp images courtesy of Josef Schlach).

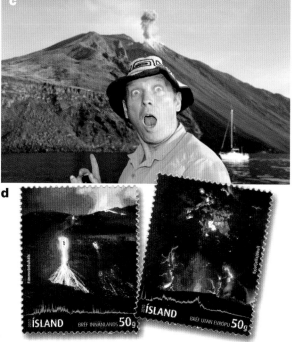

1 The world of volcanoes

Ask any child to draw a **volcano** and they will eagerly beaver away to produce an array of classic depictions of **volcanic eruptions**, volcanic islands, and towering conical mountains. Volcanoes are a keen topic in the thoughts of many children (and adults) when it comes to art and projects, and they are forming an ever more present feature in our modern lives (Figure 1.1). Our fascination with volcanoes is like some primordial link between ourselves and our planet. But what of the world of volcanoes? Many of us will not have seen an active volcano in the flesh and will never get to see one erupt. Until the Icelandic volcanic crisis (2010), which closed down airspace and disrupted everyday lives in Europe and North America, and the more recent tragedies at Mount Ontake (Japan 2014) and White Island (New Zealand 2019), our fascination with volcanoes was still that of a child: for a mythical land of volcanoes, dinosaurs and ancient Earth, somehow linked with Man. Now with an ever-increasing media profile and awareness of volcanoes, it is helpful to delve into the world of volcanoes with this guide to hot rocks.

Today there are more than 500 active volcanoes on this planet; about twenty will be erupting whilst you read this passage, and others that we can observe elsewhere in the Solar System. The Earth's volcanoes are testament to the planet's cooling and the convection of heat manifest on the Earth's surface in plate tectonics. Volcanoes allow the planet to breathe; the gases they emit have helped to develop the Earth's atmosphere and contribute a key part of the present atmospheric cycle. To understand the world of volcanoes we need to look at the Earth system as a whole, from its very centre to the volcanic plumes in the air (Figure 1.2), and the glowing eruptions and **lava** flows on the Earth's surface (Figure 1.3). We need to understand the plumbing systems that feed the volcanoes; the varied types of eruption that are seen on Earth; and to look to those beyond our planet. As volcanologists, our goal is to be able to put together ALL the pieces, from what makes rocks melt, to the different compositions of magma that result in the wide variety of eruption types and settings.

Studying volcanoes

The way in which volcanoes are studied can be placed into two main categories, which have some overlap in time and space. First, there are modern volcanoes, those that are active and erupting on the Earth's surface today. These can be monitored and measured (Figure 1.4a and b); their changes through time are noted, as are their locations on the planet, particularly where they may affect Man. Secondly, the ancient rock record is studied to see how volcanoes erupted and occurred in the past (Figure 1.4c) – how they have helped build the planet and change the course of the Earth's history. Volcanoes can

Figure 1.2 Example of a volcanic plume (eruption). 22 July 1980 eruption of Mount St Helens, second eruptive pulse as seen from the south (USGS).

be linked to the emergence of life and also to mass extinctions that have occurred at key points in our planet's history. They are important sources of materials and resources, from diamonds to the 'burnt umber' watercolours children use to paint volcanoes. The study of modern and ancient volcanoes overlaps as, with the investigation of the previous products of active volcanoes, deposits can range in age from very recent eruptions to those that are tens, thousands or millions of years old. This information from the volcanoes' recent and, sometimes, distant history helps us unravel what has happened in the past; to look for patterns of volcanic activity through time; and to use this as a form of remote monitoring to help predict what any particular volcano may do in the future.

Each major volcanic eruption that is witnessed has its own tale to tell about the way

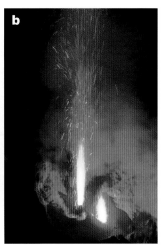

Figure 1.3 Glowing lava flows and eruptions. a) Eruption and lava flow down the snowy flanks of Mount Etna (© Wead/ Shutterstock). b) Fire fountains like Roman candles erupting at night on Stromboli volcano – a volcano famous for its regular 'Strombolian' eruptions.

in which volcanoes behave. Eruptions have an increasing effect on Man as mankind's population grows and the planet is used in more sophisticated ways, for example the impact on air travel of the eruption of Eyjaf-jallajökull in Iceland in 2010. Further, as time goes by, our chances of seeing the larger, rarer, eruptions slowly increase as the odds against them erupting reduce.

How do I start?

You do not have to travel too far to be able to start on your quest to become a volcan-ologist. Some basic background knowledge, some simple but effective field gear and a keen eye for rocks should set you on the right path to delve into the world of volcanoes.

A useful approach is to take a top-down look at the volcano or volcanic/igneous rocks that you are interested in. You can con-sider the basic components that make up the volcanic system, from its underground plumbing to the products that it erupts at the surface (Figure 1.5). Within the plumbing systems of volcanoes are magma chambers, which act as storage vats for the molten rocks on the way to the surface, and dykes and sills, which link between chambers and also from its depths to the visible volcano at the surface. Clearly in modern systems these are hidden beneath the volcano, but their movements may be monitored remotely (Chapter 9). In ancient examples, where erosion has exposed the inner workings of the volcano, the 'fossilised' magma plumb-ing system can be explored to help us under-stand how it works (Chapter 7). At the surface there may be lava flows and explosive erup-tions that lead to a wide variety of volcano types and sizes (Chapter 4). Different types of lava flows, from recent to ancient (Chapter 5) may be looked at, and modern explosive eruptions can be compared with similar eruptions preserved in the ancient rock

Figure 1.4 Sampling lava and looking at ancient deposits. **a)** Sampling lava on Hawaii, **b)** surveying a volcanic fissure in Ethiopia, and **c)** looking at pyroclastic deposits on Santorini.

record (Chapter 6). In order to better understand the products from volcanoes and their plumbing systems, a much closer look is needed at the fine scale detail shown by their associated minerals and rocks.

Igneous minerals

On the detailed scale, the rocks themselves must be understood to reveal what they mean in terms of the different volcanoes. A good start here is to understand some of the basic mineralogy of the rocks that are formed from volcanoes and their associated

'plumbing' systems. Igneous rocks are made up of various elements which, depending on the composition, lead to different crystals forming as they cool, resulting, in turn, in different types of igneous rocks. A lot can be deduced about the origin of the magma associated with a particular volcano, depending on the composition of the crystals and minerals that form as it cools. This subject will be covered in more detail in the next chapter. Close investigation into the literature of mineralogy reveals a bewildering array of different types of minerals

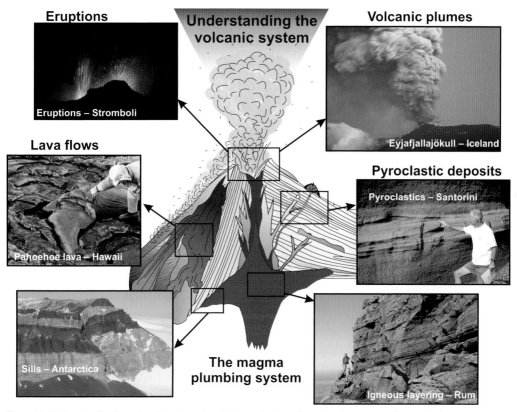

Eruptions

Eruptions – Stromboli

Understanding the volcanic system

Volcanic plumes

Eyjafjallajökull – Iceland

Lava flows

Pahoehoe lava – Hawaii

Pyroclastic deposits

Pyroclastics – Santorini

The magma plumbing system

Sills – Antarctica

Igneous layering – Rum

Figure 1.5 Understanding the volcanic system. In order to understand the volcanic system we have to look at every aspect from the volcano's plumbing system to the wide range of products it erupts at the surface (Iceland photo by Jón Viðar Sigurðsson).

that exist. Fortunately, with igneous rocks you will initially only need to master a handful of key rock-forming minerals in order to start to understand your volcano. Minerals are characterised in the following ways (Table 1.1):

Colour: the colour of the mineral under normal light; this can be very typical for some minerals, and quite variable in others.

Streak: a property of colour when a mineral is scratched along a porcelain plate (really only of use in a few examples – mainly iron oxides).

Lustre: the way in which minerals reflect light; for example, are they glassy, metallic, or dull.

Hardness: this is a scale from 1 to 10 where 10 is diamond, the hardest mineral known to us on the planet (Figure 1.6).

Cleavage and Fracture: cleavage is where a mineral breaks along planes of weakness in its structure. Minerals can have

Table 1.1 Key igneous minerals (see also Bowen's reaction series, Chapter 2).	Mineral (Typical chemical formula)	Colour
	Olivine (Mg,Fe)$_2$SiO$_4$	Olive green, yellow-green, sometimes brown
	Pyroxene (i) (Mg,Fe,Ca)$_2$Si$_2$O$_6$ (augite, etc.) (ii) NaFeSi$_2$O$_6$ (aegirine)	Black to dark green or brown to yellowish-green
	Amphibole (i) Ca$_2$(Mg,Fe)$_5$Si$_8$O$_{22}$ (OH)$_2$ (e.g. tremolite) (ii) Na$_2$Fe$_3^{2+}$Fe$_2^{3+}$Si$_8$ O$_{22}$(OH)$_2$ (riebeckite)	Black to brownish black or dark green to dark blue
	Biotite (mica) K(Mg,Fe)$_3$(AlSi$_3$O$_{10}$) (OH)$_2$	Black to dark brown or green
	Quartz SiO$_2$	Colourless to pale grey when surrounded by dark minerals; transparent
	Alkali feldspar (K,Na)AlSi$_3$O$_8$	White or pink, sometimes orange or yellow
	Plagioclase feldspar NaAlSi$_3$O$_8$ to CaAl$_2$Si$_2$O$_8$	White or green, rarely pink or black
	Muscovite (mica) KAl$_2$(AlSi$_3$O$_{10}$)(OH)$_2$	Colourless to pale brown or green

Cleavage	Lustre	Habit	Hard-ness
Very poor, usually fractures	Glassy when fresh, vitreous when altered	Usually rounded anhedral crystals, occasionally equidimensional tabular forms	6–7
2 good sets meeting at 87°/93°	Vitreous when fresh, dull when altered	4- or 8-sided prismatic crystals occasionally show-ing cleavage or aegirine more acicular	6
2 good sets meeting at 56°/124°	Vitreous when fresh, dull when altered	Prismatic or lozenge-shaped crystals often showing cleavage or riebeckite more acicular	5–6
1 excellent cleavage; cleaves into thin flexible sheets	Very shiny	Thin tabular crystals, occasionally 6-sided, espe-cially in ignimbrites and acid lavas	2.5–3
None; irregular, or curved fracture surfaces	Glassy, shiny	Rare trigonal pyramids but usually irregular, anhedral	7
2 sets at 90° poorly visible	Usually dull, some-times silky or vitreous	Tabular crystals; shiny cleavage surfaces may show simple twins. Elongate rectangular 'laths', lamellae, or irregular masses of plagioclase may be noted, in which case the crystal is termed a perthite	6
2 sets almost at 90°, poorly visible	Usually dull, some-times silky or vitreous	Lath-shaped crystals; shiny cleavage surfaces may show multiple, parallel twins	6–6.5
1 excellent cleavage, cleaves into thin flexible sheets	Shiny, silver and pearly	Tabular crystals sometimes 6-sided, especially in pegmatites	2–2.5

MINERAL	MOHS RELATIVE HARDNESS	COMMON OBJECTS
Talc	1	
Gypsum	2	Finger nail (2.5)
Calcite	3	Copper coin (3.5)
Fluorite	4	
Apatite	5	Steel nail (5.5)
Orthoclase	6	Glass plate (6)
Quartz	7	Streak plate (7.5)
Topaz	8	
Corundum	9	
Diamond	10	

Figure 1.6 Mohs hardness scale. Ranges from 1 (softest) to 10 (hardest) with key minerals for each hardness range and the hardness of some common objects indicated.

one, two or three cleavages. Fracture is the way in which a mineral breaks apart in a non-linear fashion and can be characteristic of some key minerals.

Shape (also known as habit): minerals can be described as being euhedral (good crystal form), subhedral (moderate crystal form) and anhedral (showing no crystal form). The crystals themselves may be bladed, tabular, fibrous, blocky, rounded, or prismatic.

Other properties: including taste (for example, halite – rock salt), magnetism (for example, magnetite).

In most examples you will not be able to test or see all of these properties, but they are useful to know in case you have large crystals in your volcanic rocks.

Additionally, minerals can be subdivided into seven **crystal systems**, based on their symmetry and physical properties: cubic, tetragonal, orthorhombic, monoclinic, triclinic, trigonal, and hexagonal. Again, a detailed understanding of these systems is not important initially, but when faced with a table of mineral types you will inevitably come across these names. As you get to

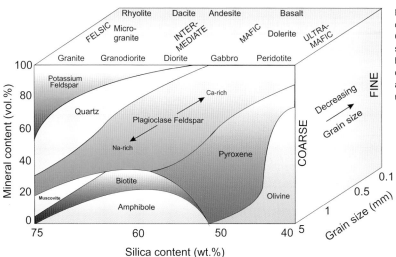

Figure 1.7 Igneous classification scheme. General classification scheme for igneous rocks based on grain size, silica content and relative abundance of igneous rock-forming minerals.

know your minerals better, you will start to see how things like the crystal system will help dictate the form of the resultant minerals, and why they have certain key properties. Some of the key minerals and their characteristics are given in Table 1.1. As the discussion develops in Chapter 2 concerning how rocks melt, it will become clear how these minerals fit into our understanding of volcanoes, and the way in which they can be used to help constrain volcanic process.

Basic classification of igneous rocks

We do need to know the basic classification system of igneous rocks – how they are grouped in order to put each type of volcanic rock into the broader context. **Igneous** rocks are classified according to the different relative amounts of the key minerals outlined in

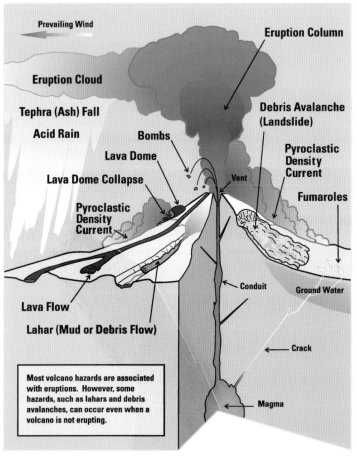

Figure 1.8 Summary of volcanic hazards (Adapted from USGS).

Prevailing Wind

Eruption Column

Eruption Cloud

Tephra (Ash) Fall

Acid Rain

Bombs

Lava Dome

Lava Dome Collapse

Pyroclastic Density Current

Debris Avalanche (Landslide)

Pyroclastic Density Current

Vent

Fumaroles

Lava Flow

Lahar (Mud or Debris Flow)

Conduit

Ground Water

Crack

Magma

Most volcano hazards are associated with eruptions. However, some hazards, such as lahars and debris avalanches, can occur even when a volcano is not erupting.

Table 1.1. Each mineral is made up of its constituent elements, and as the quantity of each mineral type varies, different rocks are composed. In general, igneous rocks range in composition from being relatively rich in iron (Fe) and magnesium (Mg), known as **mafic** rocks, to rocks that are rich in **silica** (SiO_2, silicon oxide) and aluminium (Al), known as **acidic** rocks.

Rocks can be classified by their **grain size** as well as by their chemistry. This is used to help distinguish **plutonic** rocks, those which have cooled slowly at depth, from shallow intrusions and **volcanic** rocks, which erupt at the Earth's surface and cool very quickly. With plutonic rocks, the slow crystallisation leads to large crystals, whereas rapid cooling results in very small-grained crystals or even glass. A simple classification scheme for igneous rocks is presented in Figure 1.7, which combines the relative compositions of the rock types with the grain size to give a number of rock classification names. Basic rocks rich in iron and magnesium range from coarse-grained **gabbros** through medium-grained **dolerite** to fine-grained **basalt**. You may come across people calling mafic rocks 'basaltic' in composition, which can sometimes be used in a non-grain-size sense. Acidic rocks range from coarse-grained **granites**, through **microgranites** to fine-grained **rhyolite**. Intermediate between these are **andesites** and **dacites** (Figure 1.7).

Volcanoes as hazards

Active volcanoes present a number of hazards to Man. Due to the very fertile land around volcanoes, Man has long colonised many of the world's volcanically active areas. Indeed, volcanic island chains and isolated island hot spots owe their very existence to the volcanoes that build them. There are the direct hazards associated with the eruption of a volcano. These, as shall be demonstrated, are particularly important for some of the more explosive volcanoes (summary in Figure 1.8). **Pyroclastic Density Currents** (previously termed '**pyroclastic flows**'), bombs, ash and lava direct from a volcanic eruption can have devastating effects on the immediate and closely surrounding areas. Associated with any direct explosive activity may be the collapse of part of the side of the volcano or the collapse of **domes** of new magma that collect at the top of the volcano. Further away, an eruption can destabilise the landscape, leading to mudslides (**lahars**) which can often be more deadly than the original eruption. For example, a mud flow in Columbia, after the 1985 eruption of Nevado del Ruiz stratovolcano in Tolima, is thought to have killed some 23 000 people. Even more remote are the effects on the climate from large eruptions, which can change the planet's climate for a few years, as has been experienced during Man's time on the planet, or, in extreme cases, change the whole course of evolution.

In order to fully understand our volcanic planet it is necessary to understand why rocks melt; to understand how and where volcanoes occur on our planet; and to recognise the types of volcano and why they occur. The differing sizes of volcanoes and how they are classified should be learnt. The past is the key to the present, and so a good understanding of key volcanic events through time also helps introduce the world of hot rocks.

2 The cooling Earth – how do rocks melt?

For the Earth to have volcanoes there must be a source of heat within the planet, there must be a mechanism for melting rock, and there must be a way for that molten rock to get to the surface. To understand why rocks melt, the starting point is the current structure of the Earth. Our planet is made up of layers that are broadly defined by their composition and behaviour. Figure 2.1 provides a schematic cross section through the Earth, from its centre (some 6370 km from the surface) through to the thin veneer of crust and atmosphere.

The Earth's basic structure can be broken down into the following layers: the **inner core** with a radius of some 1200 km, which is solid and almost entirely composed of iron. The outer core surrounding this is about 2300 km thick, is liquid, and is composed mainly of a nickel-iron alloy. The **Earth's magnetic field** is believed to be controlled by this liquid outer core. The next layer from the centre is the **mantle**, which is around 2900 km thick, is so viscous as to be almost solid, and is composed mainly of silicates rich in iron and magnesium. The slow circulation of heat

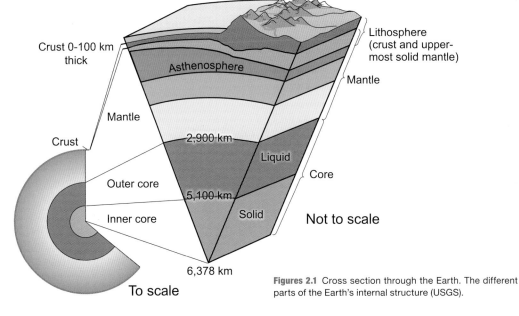

Figures 2.1 Cross section through the Earth. The different parts of the Earth's internal structure (USGS).

through the mantle over time is thought to be one of the driving forces for plate tectonic processes, and is important in our understanding of volcanoes. The outermost layer of the Earth is the crust, which can be divided into continental and oceanic crust. The crust has a variable thickness, with **continental crust** being some 35–70 km thick and the **oceanic crust** some 8–10 km thick. The continental crust is composed mainly of **silicates** rich in aluminium, whilst the oceanic crust contains more iron and magnesium. The crust and the uppermost part of the mantle, which behave as the rigid shell of the planet, are known as the **lithosphere** (*see* Chapter 3).

The Earth is believed to have begun as one molten mass with all elements mixed together. Since then, heavy and light elements have separated in a process known as **differentiation**, resulting in the central core, containing the larger amounts of heavy elements, with the other layers containing increasing percentages of lighter material the further they are from the core, until the atmosphere, made up of gases, is reached. It is as part of this ongoing differentiation process that modern volcanoes act to release gases into the atmosphere from the interior.

The role of radiation

The Earth, and therefore all that composes it, is believed to have originated as a ball of molten material around 4.6 billion years ago. Yet if the Earth had simply been cooling over 4.6 billion years up to the present, it would have long since stopped cooling and become completely solid. William Thomson, later known as **Lord Kelvin**, famed for the absolute temperature scale that bears his name,

calculated a cooling time for the Earth based solely on radiation of heat at around 30 000 years. This cooling time is clearly much shorter than the age of the Earth as it is now known, at around 4.6 billion years, so why has it not fully cooled? The answer lies in the composition of the rocks that make up the Earth. The decay of **radioactive elements** in these rocks provides an additional heat source for the planet. These elements include uranium and thorium amongst many others, which are slowly decaying at variable rates through time and releasing heat as they do so. At each stage of decay, heat is given off (an **exothermic** reaction), and with the half-life of the main uranium isotope U^{238} being ~4.5 billion years, this process can go on for a long, long time. It is due to the decay of these radioactive minerals that the Earth is still hot and able to melt rocks, and feed the volcanoes that fascinate us. With a constant internal heat source from radiation, and limited ways to lose this heat, the Earth stays warm. But the molten rock at the surface comes from the solid mantle, so just how are rocks melted?

Dr Volcano's oven guide to melting rocks

Turn your oven up to Gas mark 44 (electric 700–800 °C) if you have a granitic rock, or gas mark 74 (electric 1100–1200 °C) for basaltic rock, place in oven and wait (Figure 2.2). This simple guide to melting rocks is not too far from the truth, though one can note that the temperatures needed are much more than a household oven will reach. In order to understand at what temperature different rocks melt and at what temperatures different crystals will start to solidify from molten rocks, a number of simple, yet highly

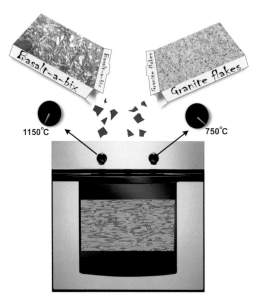

revealing, melting experiments were needed. A classic study in the early 1900s by **Norman L. Bowen** (the head chef of the igneous world) created the most important recipe book for igneous geology and volcanology, known as **Bowen's reaction series** (Figure 2.3). He conducted a number of melting experiments, taking a variety of compositions from basalts (rich in iron and magnesium) to granites (rich in aluminium and silicon) and those in between (*see* Figure 1.6). These experiments showed that basaltic compositions melt at much higher temperatures, and granitic rocks at a much lower temperature.

Figure 2.2 Dr Volcano's guide to melting rocks. Granitic (silicic/felsic) composition rocks melt at lower temperatures than basaltic (mafic) composition rocks.

Figure 2.3 Bowen's reaction series. Key igneous rock-forming minerals and how they relate to melting temperature and rock composition.

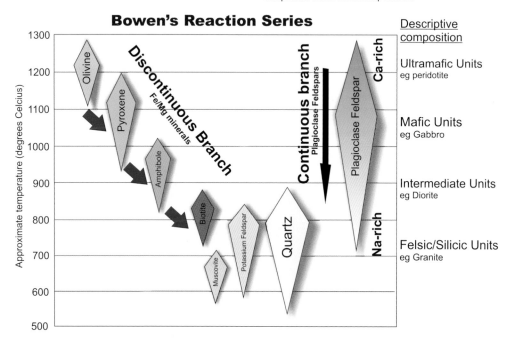

Bowen's experiments also showed the different minerals that crystallised at different temperatures (Figure 2.3). Thus, minerals such as olivine, pyroxene and calcium (Ca) rich plagioclase crystallise from hotter, more mafic compositions, for example basalt. Minerals such as quartz, potassium (K)-feldspar, and micas crystallise from cooler, acidic compositions such as granite; and amphibole and more sodium (Na) rich plagioclase are typical of intermediate compositions (Figures 2.3. and 1.6). Once minerals in a volcanic rock are identified, inferences may be made as to the melting temperature and thus the magma temperature that formed it. Bowen's experiments still underpin much of our understanding of melting in igneous systems.

Melting experiments are still of great importance to us. For example, in Figure 2.4, recent melting experiments demonstrate the correlation of interlocking crystal frameworks with distance away from a contact in the Palisades Sill, USA. Cubes of rock are subjected to melting conditions for set periods of time. The rocks closest to the contact (0.3 m) completely melt straight away, showing that there is little crystalline structure. As the melt cubes get further into the sill, only a partial melt occurs, with more and more of a crystalline structure, with the cube of rock maintaining its shape and just a small trickle of melt seeping out of the base (3rd and 4th cube, Figure 2.4). By looking at the melting and crystallising behaviour of rocks under different conditions and with different compositions, much may be learnt about possible conditions within the Earth's interior, and also how lava behaves at the Earth's surface. The 'Lava Project' at Syracuse University, USA, uses molten rocks to explore how lava flows under different conditions and over different substrates (e.g. Figure 2.5), and also bridges into art and education using melt experiments. With improved methods of investigating the melting of rocks under differing pressures, their melting at different depths can be explored. Melting experiments

Figure 2.4 Modern melt experiments. A series of melt experiments at progressive distances from the contact zone of the Palisades Sill, USA. The example closest to the contact (0.3 m) completely melts straight away, showing that there is little crystalline structure. As the melt cubes get further into the sill, only a partial melt occurs, with more and more of a crystalline structure, with the cube of rock maintaining its shape and just a small trickle of melt seeping out of the base (e.g. 14 m and 21 m) (Lori Dickson).

| 0.3 m | 6 m | 14 m | 21 m |

Distance from sill contact

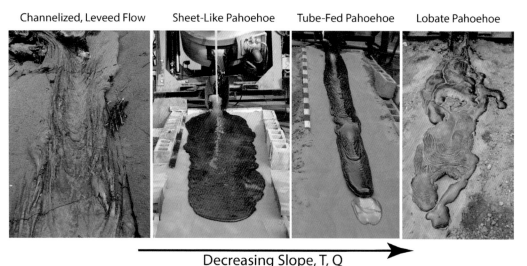

Channelized, Leveed Flow   Sheet-Like Pahoehoe   Tube-Fed Pahoehoe   Lobate Pahoehoe

Decreasing Slope, T, Q

Figure 2.5 Lava flow morphologies correlated with lava temperature, slope and effusion rate (Q), from experiments using re-melted Precambrian basalt poured over dry sand. Similar morphologies can be seen in natural examples (Syracuse University Lava Project (http://lavaproject.syr.edu) J.A. Karson & R.J. Wysocki, unpublished data).

on meteorites and other extra-planetary materials may be performed to help understand their formation temperatures in the Solar System. Ultimately this information may be used to understand how rocks melt to make volcanoes.

Melting the Earth

With such rock experiments, the ways in which different rock compositions melt at different temperatures can be learnt; but how rocks melt on the Earth today still has to be understood. Drilling a core section of the Earth from the surface down shows that the temperature and the pressure conditions increase with depth from the surface. The rate at which the temperature increases with depth is known as the **geothermal gradient**, and commonly ranges from 25–30 °C per kilometre depth in most parts of the world.

Exceptions to this are areas of thin crust/hot spots with very high geothermal gradients (30–50 °C/km) and thickened crust in subduction settings (5–10 °C).

Following the geothermal gradient alone, the rock-melting temperatures learnt from the experiments above will soon be reached. Why, then, is the whole of the Earth below a certain depth not all molten magma? This is where pressure becomes an important factor. The higher the pressure is, then the higher the temperature must be to melt rocks. This simple relationship means that for most circumstances on Earth, the pressure/temperature relationships are such that the rocks are below their melting point, and not hot enough to melt. A simple plot of an average geothermal gradient is shown in Figure 2.6. Also on this plot is a line known as the **solidus**. This links the temperature/

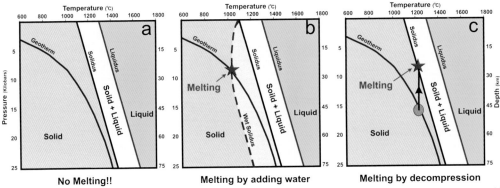

Figure 2.6 The Earth's geotherms and melting: **a)** Normal geotherm relationship with the solidus and liquidus lines; **b)** addition of water lowering the solidus and allowing melting; and **c)** rapid decompression of the mantle allowing melting. (Note: above the solidus the rocks start to melt and will contain solid and liquid. Above the liquidus the rock is completely molten and will contain only liquid).

pressure relationship to the depth at which melting will occur. Another line also appears, called the **liquidus**. This is the point at which the whole of the rock would be a liquid. Between the solidus and liquidus, molten material and solid material will co-exist. As you start melting at the solidus, you will initially have a small amount of melt and a large amount of solid. As you move towards the liquidus, the volume proportion of melt increases and of solid decreases, until you reach completely liquid when you cross the liquidus line. Our experiments have demonstrated that different rock compositions melt at different temperatures. The example shown is for a rock of basaltic composition, but there appears to be a problem: as you can see from the diagram (Figure 2.6a), the two lines do not cross, and so no rocks should melt. So how are there basalt melts at the surface of the Earth, such as those we see at Hawaii and Iceland and many, many other places?

The answer lies in two simple but very effective mechanisms. The first involves water and its relationships to the melting temperature of rocks. If water is added to rocks, their melting temperature will drop. This has the effect of moving the solidus line towards the geothermal gradient: if they cross, the rocks start to melt (Figure 2.6b). Thus a new line known as a **wet-solidus** can be plotted; its exact position will change depending on the amount of water present. So with rocks at a certain temperature and pressure, once water is added to them, they will start to melt. The other method to get rocks to melt is related to releasing the pressure on them. If the pressure is quickly removed from a rock, faster than it is allowed to cool, the geotherm will be raised, as the rock will be hotter in relation to the pressure expected at a shallower depth (Figure 2.6c). Thus, on Earth, rocks will melt either when water is added or when the pressure on them is released through **decompression**.

Finally, we cannot ignore the role of time and heat supply. It is possible to melt the surface of a small rock with a blow torch

Figure 2.7 The author performing rock heating experiments with Mount Etna in the background.

(Figure 2.7), but it would take a full furnace many hours to melt the rocks used in the experiments highlighted in figures 2.4 and 2.5. In the case of the Earth, with its giant radioactive heat engine, even then we need the ideal circumstances of tectonic setting and timing in order to create our hot molten rock – magma.

Origins of magma

So rocks will melt, both experimentally and within the Earth; but where does magma get generated in the Earth and where does it come from? A number of markers can be used to tell us at what depths a magma that reaches the surface was made. We can use remote studies like **seismic signals** to locate where melt is moving at depth (*see* Chapter 9) and also to pinpoint areas where melt may be being generated, as certain seismic signals are slowed through partially molten zones, and do not travel through liquids. Volcanoes themselves provide the window into the inner Earth, and it is from their products (lava, pumice and ash), that the origins of

magma are understood. Molten lava erupted at the surface often contains crystals that have grown at certain temperatures and pressures. The lavas may also contain rock **fragments**, known as **xenoliths**, which they have dragged up from depth. Such xenoliths can provide key information about the depth the magma has travelled, and about its pathway to the surface. The chemical composition of the lava can provide clues as to the depth of formation, and the types and composition of gases given off by the volcano can also be used to say something about melting depth. Some of the magmas from the greatest depths known are found at unusual volcanoes called **kimberlites**, after the town of Kimberley in South Africa. These volcanoes are special because they can contain diamonds (Chapter 7) and other minerals formed at high pressure. These minerals signify great depth of formation, between 150 and 450 kilometres down.

Looking at more common, shallower magmas, it can be a little harder to ascertain their melting depth without key minerals, such as diamond, to guide us. This is often where the humble crystal that is growing from the magma can provide a surprising array of information about the magma in which it is created. Looking at the insides of these crystals reveals how they have grown through time. This can be demonstrated by zones of different chemistry from the centre (core) of the crystal to its outsides (rim), similar to the growth rings of trees. As the growth rings of a tree tell us about the environment in which the tree grew, so too, the zones in a crystal tell us about the magma conditions during its growth (Figure 2.8).

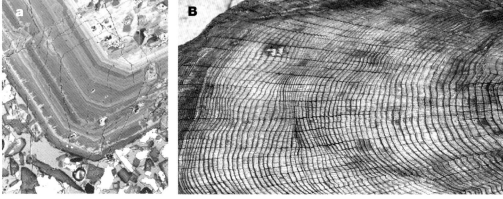

Figure 2.8 Zoning in crystals vs zoning in a tree. The zoning in a crystal is similar to the zoning in a tree shown by its tree rings. Each progressive growth records the environment at that time, and valuable information about the past can be locked up in this information (Photo **a)** Jon Davidson, **b)** Adrian Pingstone).

First, the largest crystals within the erupted rock are looked at, as they are the ones that will have started growing the earliest in the magma, at depth. These are used to help build up a picture of the magma's origin as it has risen up through the crust. Crystals can also trap droplets of melt on their way to the surface. These **melt inclusions** provide specific examples of the early magma, and the melt inclusion chemistry can also be used to help to constrain depth of melting. Using such data, most melts are revealed as forming in the Earth at between 50 and 200 kilometres depth. This is relatively shallow, given the scale of the planet, and is intimately related to the plate tectonic system in operation at the Earth's surface.

To summarise this story of melting rocks: temperatures in excess of >700 °C are required for granitic compositions, or about 1100 °C for basaltic ones, to melt rocks. Melting occurs usually at depths of between 50 and 200 km, and either fluids (e.g. water) need to be added, or there has to be a release of pressure, or a combination of both, for them to melt.

3 Volcanoes, plate tectonics and planets

Our planet is unique in our solar system with its system of 'plate tectonics'. Plate tectonics are a large factor in the location of volcanoes on the planet. Alfred Wegener noted in 1915 that many of the continents on the Earth today could fit together like a jigsaw. The matching shapes of South America and Africa are one very striking example. Wegener proposed that these continents were once joined together and that they had drifted apart through 'continental drift'. This proposition was not well received by scientists at the time, particularly because no explanation could be given for how the continents might move.

Later, key work by Arthur Holmes hypothesised that the Earth was much hotter than originally thought, and that the Earth's mantle could deform in a ductile fashion, very slowly over thousands of years. The main problem remained that no driving force for such a process could be identified, and the evidence linking the continents was not thought strong enough to fully support such a model. Palaeomagnetism provided the clue. As molten rock cools, the iron minerals in it take on the orientation of the magnetic field of the Earth at that time. Old volcanic rocks can thus be used to map out the Earth's magnetic field and the relative position of parts of the Earth to that field, and how they have varied through time. The final steps to understanding this continental jigsaw puzzle were made in the 1960s, when palaeomagnetic data was used to show the positions of the magnetic poles through Earth's history in relation to the rocks on the continents. Around the same time the magnetic polarisation patterns (magnetic stripes) on the sea floor were also imaged to show how the ocean crust is developed through time. These palaeomagnetic data showed that the continents must indeed have been together in the past. From this, the theory of plate tectonics has been developed to show how the continents are moved around the planet by convection. In order to conserve mass so that the Earth maintains a constant size, new crust is developed at constructive plate boundaries and older crust is recycled at destructive plate boundaries. One can construct a map of the Earth that shows how these plates are configured (Figure 3.1a). These plates and the forces that drive them play a pivotal role in our understanding of volcanoes.

Plates and melting

The plate tectonic model has completely revolutionised the study of geology. For the first time the way in which the continents have moved over time may be mapped to show the geology of the world through time. This helps us understand many modern phenomena such as earthquakes and volcanoes. Overlaying the locations of the world's recent earthquakes and active volcanoes on a map of the Earth's continental plates demonstrates a very remarkable correlation (Figure

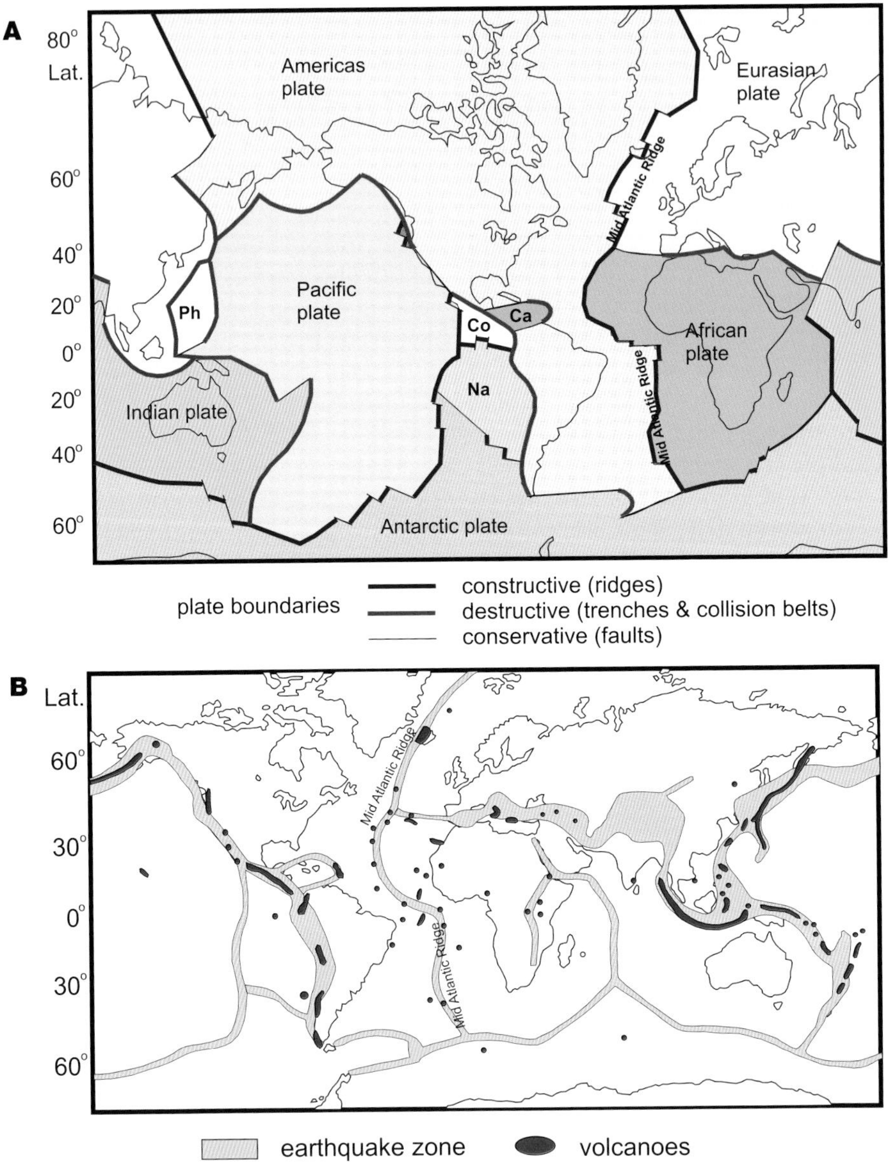

Figure 3.1 Plate tectonic maps highlighting volcanoes and earthquakes. **a)** Map of the main plates and their boundaries (Na – Nazca; Co – Cocos; Ca – Caribbean; Ph – Philippine). **b)** Distribution of earthquakes and recent volcanoes. This highlights that the earthquake activity and many of the volcanoes are located in and around the plate boundaries (from Graham Park, 2010).

3.1b). All of the major earthquakes and the great majority of volcanoes (we shall look at the exceptions later) are found on the plate boundaries. This is no coincidence, and indeed, material is fed to volcanoes at plate boundaries through plate tectonic processes.

Volcanoes are fed by the plate tectonic engine, which drives rocks into situations where they start to melt. That melt then makes its way to the surface to form volcanoes because it has low density and is relatively buoyant. To get molten rock at the Earth's surface, either the pressure on a hot rock needs to be released relatively quickly, before it can cool, or water needs to be added to a hot rock in order to get it to start melting (Figure 2.5, Chapter 2). Plate tectonics provides the perfect examples of both these mechanisms, as is shown by two key types of plate margins for volcanic activity: constructive and destructive; with very few examples of volcanoes being found at strike-slip/conservative margins (*see* Chapter 5 of *Introducing Geology* in this series for a fuller description of how continents move).

Melting at constructive plate margins

The simplest of the melting mechanisms occurs at **constructive plate margins**; 'constructive' because new crust is being created on the Earth's surface as two plates are moving away from each other in opposite directions. Each acts as a giant conveyor belt of crust and upper mantle, known as the lithosphere, which acts as the solid outer shell of the planet. As the plates separate, the Earth's mantle is drawn upwards towards the ridge that marks the plate boundary (Figure 3.2). This general conveyor motion of the separating plates provides a decompression

mechanism whereby the hot mantle below is drawn upwards into the gap. This releases the pressure on it and creates partial melts via **decompression melting** (cf. Figure 2.5, Chapter 2). Partial melts are small percentages of melts, as the rocks are just above the solidus but below the liquidus, in the region where the rock will still be mostly solid. As you move away from the solidus and towards the liquidus, the amount of melt generated increases (*see* Chapter 2). These partial melts find their way to the ridge axis, where they form new ocean crust by intrusion and eruption on the sea floor as volcanoes. This process is repeated time and time again. Thus constructive plate margins provide the new ocean crust on the planet. Volcanoes occur all along the ridges that make up these constructive margins, for instance at the **Mid-Atlantic Ridge** (Figure 3.1). For the most part these margins are under water in the deep oceans and are rarely seen.

A constructive margin can be witnessed on Iceland, where the Mid-Atlantic Ridge rises above the ocean. Iceland straddles the European and American plates, with a ridge system that completely dissects the island (Figure 3.3). A person can place one foot in America and one foot in Europe, hopping from one continental plate to another, or even dive between plates (Figure 3.3c). The plates are separating at the rate that our fingernails grow. On land many eruptions of volcanoes occur along these **fissures**, where they mainly erupt **pahoehoe** type lavas and can have episodes of fire-fountaining. Volcanoes that erupt under the oceans mostly form pillow lavas (*see* Chapter 5 for an explanation of lava types). Another area on the planet where volcanic activity associated

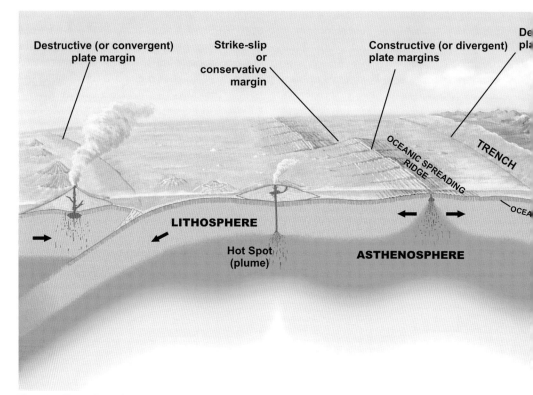

Figure 3.2 Three-dimensional plate boundaries model. The three types of plate boundary/margin found on Earth: a constructive (or divergent) plate margin; a destructive (or convergent) plate margin; a strike-slip or conservative margin (USGS).

with constructive plate margins can be seen is in the **Afar Rift** in Ethiopia, Africa. Here the early stages of rifting are observed, which if continued could lead to the separation of parts of the African continent, forming a new constructive plate margin.

Melting at destructive plate margins

Where two plates converge on each other and collide, one plate is **subducted** (descends) beneath the other. This happens with oceanic crust being subducted under continental crust. Such a plate margin is known as a

destructive (or **convergent**) **plate margin** (Figure 3.2). Here the forces are mainly compressional, and a process of **crustal thickening** occurs, such that it seems difficult to create melt through a decompression process like those seen at constructive margins. However, the subducting plate (lithosphere), which is driven under the crust and back down into the mantle, known as the **subducting slab**, carries a carapace of oceanic crust with it. This piece of oceanic crust contains water within its minerals and associated sediments. So there is the possibility of water

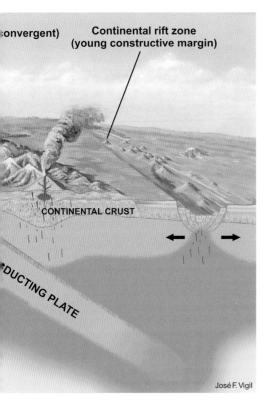

**Continental rift zone
(young constructive margin)**

:onvergent)

CONTINENTAL CRUST

:DUCTING PLATE

José F. Vigil

pyroxene, which does not. Such reactions drive water off the slab, and the key question is where does the water go? This reaction driving off water occurs at depths where the plate has been subducted beneath the crust and part of the mantle of the overlying plate. The waters and fluids that are driven off from the slab work their way up into this mantle stream, known as the **mantle wedge**, which contains hot rocks, primed for melting if water is added. So the addition of the released water to the hot rocks lowers their melting temperature at the pressure that they are under, and they start to melt (Figure 3.2). At **subduction zones** these melts make their way to the surface to feed volcanoes; for example, all around the Pacific Ocean, which is encircled by subduction margins, volcanoes are seen in the so-called **ring of fire** (Figure 3.2).

An interesting result of this relationship between plates and melting at destructive margins is the depth at which the water gets driven off, and the subsequent location of the volcanoes on the plate's surface. Along the line of a subduction zone, at a destructive plate margin, volcanoes occur some distance away from the subduction zone itself. That is because the plate has to be buried to some 100 km or so for the fluids to be driven off. The location of the volcanoes fed by the melting can then be used to calculate how steeply the subducting slab is sinking into the mantle. Where volcanoes are closer to the destructive margin the plate is dipping steeply; where the area of volcanism is further away the slab is at a shallow angle. For example, along the Andes there are abrupt changes in subduction dip from nearly flat beneath Colombia (3–5.5° N), to steeper in the Central Andes (18–28° S), and

being transported to hot rocks and to produce a melt by reducing the melting temperature by adding fluids (cf. Figure 2.5, Chapter 2).

At first it was thought that the presence of water in the slab would cause the slab itself to melt when it reached a depth where it would heat up to a significantly high temperature, so-called **slab melting**. But it is now known that when certain pressures and temperatures are reached, reactions occur with changes from one mineral form to another to release water: for example, from amphibole, which contains water in its structure, to

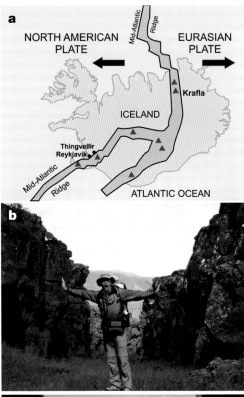

Figure 3.3 Iceland ridges map and photo of ridge. **a)** Map showing Mid-Atlantic Ridge through Iceland with key volcanoes labelled. **b)** Photo of the Mid-Atlantic Ridge at Þingvellir (Thingvellir), Iceland. **c)** Author diving the Mid-Atlantic Ridge at Lake Silfra, Iceland.

steeper again beneath south central Chile (34–42° S).

There are examples where both subduction- and rift-related volcanism occur in close proximity to each other. This can often occur where oceanic lithospere is being subducted under other ocean lithosphere, and also in areas where ocean crust is being subducted under thin continental crust. The development of the destructive plate margin leads to a roll-back mechanism whereby extension and rifting is developed on the overriding plate. This is behind the volcanic arc that marks the volcanoes originating from subduction-related melting. This extension gives rise to what is known as the back-arc basin and can involve decompression melting and volcanism in the form of an associated spreading ridge (Figure 3.2).

What about strike-slip/conservative margins?

Where two plates move past each other side to side in opposite directions, a **strike-slip** or **conservative margin** occurs. Here crust is neither created nor destroyed, and it is very rare for volcanoes to occur at such margins. The main feature of such margins are earthquakes. Strike-slip faults produce some of the most violent earthquakes. However, at conservative margins, there is never an exact or true strike-slip motion. Many have a component of oblique slip to them, and often the faults themselves are not in exactly straight lines. Where a fault meanders somewhat, it is possible to create areas of compression and areas of extension along the fault. First, this helps explain the large earthquakes, as an irregularly shaped fault will not glide easily and may lock up, with the pressure unlocking only in quick, violent releases. Where there is net compression on the fault no material is being subducted, so there should not be rock melting by fluid release, as at subduction zones. However, where there is net extension along part of the fault there may be some local uplift of the hot mantle, resulting in some decompression melting, and this may explain the rare occurrences of some volcanoes along predominantly strike-slip systems.

Exceptions to the rule – 'pluming hot spots'
There are a number of places on the planet where the location of volcanoes does not correspond to the plate tectonic map (Figure 3.1b), for example in Hawaii. These situations require specific attention in order to complete our understanding of why volcanoes occur where they do on the Earth. The mechanisms of melting that we discussed in Chapter 2 are still required, so an explanation is needed as to why melting occurs 'within plate', i.e. not associated with plate margins, where the addition of water or decompression causes melting from the plate tectonic processes.

When it comes to within-plate volcanism, you cannot find a place much further away from a plate margin than Hawaii and its associated chain of islands. In the middle of the Pacific plate, it is literally thousands of miles from any plate-tectonic driven melting, yet it is one of the most famous volcanic areas on the planet. Volcanism has been occurring where Hawaii is for millions of years, which is demonstrated by the chain of volcanic islands and underwater **seamounts** that extend away from it. As you move along this chain, the volcanic islands and seamounts get progressively older the further they are away from the sites of present day volcanism. The chain

extends in the same direction as that in which the Pacific plate is moving: it is shifting towards the north-west at a rate of ~7 cm/year. (Figure 3.4). So the volcanic activity stays in one place whilst the plate moves, producing a chain of islands. This feature is also common to a number of other within-plate volcanic zones, such as Yellowstone in the USA.

This relationship of static volcanic zones with drifting plates can only mean one thing: that the melt generating process is deeper than the plates. Looking at the volcanoes themselves it will be noted that they produce some of the hottest lavas (temperatures in excess of 1100 °C) to be found on the planet today. The phrase '**hot spots**' has been coined to describe this relationship of the volcanic zones on the planet that do not move in relation to the plate tectonic engine. Plumes of anomalously hot mantle material produced as part of the Earth's convective cooling are thought to be their cause. Volcanoes form where the hot mantle is rising upwards in one of these plumes and producing **decompression melting** as the hot material rises faster than it cools, thus providing a mechanism for producing melt driven by processes deeper in the Earth than plate-driven motions. Sometimes these hot spots are called '**mantle plumes**', and there is some debate as to where they originate from within the planet. Do they come from the core/mantle boundary or higher within the mantle? Whatever their origin, these plumes of anomalously hot mantle are capable of generating melting and volcanoes on the surface anywhere on the planet.

Where a hot spot or plume intersects a plate boundary there will be more melting potential than would be expected from one

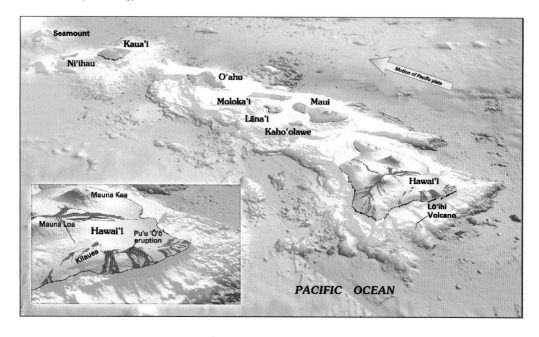

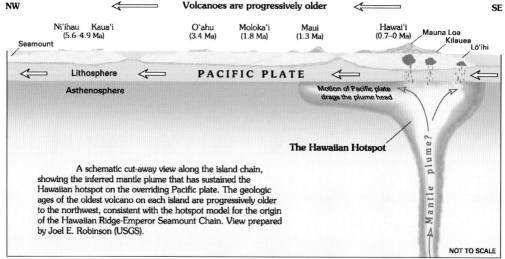

Figure 3.4 Hawaii island chain and hot spot. Schematic highlighting the trail of Hawaiian Islands produced by the Pacific Plate moving over the Hawaiian 'Hot Spot' (USGS).

mechanism alone. This phenomenon is best displayed at Iceland, where the constructive plate margin of the Mid-Atlantic Ridge coincides with the **Iceland Plume**. The melting that occurs there is far greater than anywhere else along the ridge. The crust created is much thicker and the buoyancy forces associated with the thick crust and the hot spot raise the surface of the lithosphere above the ocean to form land. Thus Iceland rises out of the sea at the coincidence of a hot spot and a ridge. Not only is the ridge exposed above the ocean for us to see it in great detail, but the amount of volcanism also allows us to observe many live volcanoes. Iceland was the site of one of the largest lava flows in recorded history, that of the **Laki eruption** in 1783. Where these plumes have coincided with plate boundaries in the past, there have been some massive outpourings of lavas known as **flood basalts** (e.g. the **Columbia River basalts** and the **Siberian traps**). These commonly occur where hot spots coincide with early rifting events marking the break-up of continents. Flood basalts and their effects will be considered later, in Chapter 4.

Volcanism on other planets

Although the Earth is distinct in our solar system in having plate tectonics, it is not unique in its volcanism. Many of the planets in our solar system display either ancient volcanism or even volcanoes that are erupting today. From our own Moon to moons from some of the more distant planets and on the planets themselves, for instance Mars and Venus, the planets around us can add to our introduction to volcanology. Our Moon most probably formed from a collision between the proto-Earth and a large external body some 4.5 billion years ago, during the early stages of the development of the Solar System. Although now fully cooled, the Moon formed from molten rock, and a number of features on its surface can be attributed to lavas and a volcanic origin. Vast flat areas called **mares** (Latin for sea) are likely to represent vast flows of basalt. One such example, the Mare Imbrium, is a feature that is thought to have formed when lava flooded the giant crater left by an impact. Many volcanic features can be recognised, including volcanic **vents** with associated domes and cones. Many of the other Mare surfaces represent smooth surfaces of large lava flows or giant lava lake-like features. One type of feature that has been attributed to lava, the long sinuous channels or levée-like features known as **rilles** (for example Hadley Rille, Mare Imbrium), are thought to be generated by lava flows or lava tubes, transporting lava beneath the surface. The samples from the Moon's surface collected by astronauts include predominantly basalts, coarser gabbros and partly welded breccias. The original volcanic features on the Moon are somewhat masked by millions of years of impact craters, and care must be taken to distinguish true volcanic craters from impact related craters.

Venus has a similar size to Earth, and its surface was mapped in detail by the NASA's Magellan mission in the early 1990s. The resultant images have revealed a wonderful array of volcanoes and volcanic forms including collapse calderas, volcanic centres, dome-like structures and linear features that may represent dykes (these volcanic features will be explained in detail in Chapters 4–6). The atmosphere on Venus is cloudy and it has taken this mission to fully recognise these

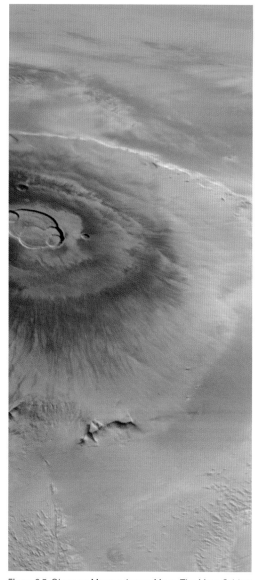

Figure 3.5 Olympus Mons volcano, Mars. The Mars Orbiter Camera obtained this spectacular wide-angle view of Olympus Mons on Mars Global Surveyor's 263rd orbit, around 10:40 p.m. PDT on April 25, 1998 (NASA/JPL/Malin Space Science Systems).

volcanic features with surface mapping. The features are not all old, as it can be shown that major outpourings of lavas, akin to our flood basalts on Earth, have re-surfaced parts of Venus and may be as young as 300–1600 million years, based on impact crater relationships.

Mars is home to one of our solar system's truly 'super' volcanoes, **Olympus Mons**, a massive shield volcano that rises some 27 km from the plains that surround it and is roughly 600 km in diameter (Figure 3.5). Perhaps the largest volcano in our solar system, Olympus Mons is the biggest of four big shield volcanoes named Mons (Latin for mountain), all of which dwarf any on the Earth. Formed by massive outpourings of lava, it may represent a large plume on Mars, but with no plate movements (as in the Hawaiian example) the material builds up in one place to form the massive shield volcano. Given Mars' smaller size compared to Earth, you can imagine these volcanoes make a big impact on its landscape, and even its central caldera complex is big by Earth standards – 85 x 65 km – and it is thought to have been active as recently as 2–115 Ma ago (based on impact crater evidence) and so may still have a part to play in future volcanic eruptions. The surface of Mars also shows many volcanic features that can be related to structures we see on the Earth's surface (*see* Chapter 6, Figure 6.11).

So far all of the planetary volcanoes mentioned erupted some time ago, but what of contemporary volcanism outside of Earth? On one of Jupiter's moons, 588 million kilometres (365 million miles) from Earth, there are signs of active volcanism. The moon Io, one of the four Galilean satellites, erupts frequently, and fantastic images of these eruptions have been captured by the Galileo

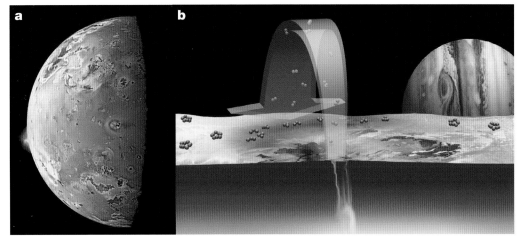

Figure 3.6 Volcanism on Jupiter's Moon Io. **a)** This colour image, acquired during Galileo's ninth orbit (C9) around Jupiter, shows two volcanic plumes on Io. One plume was captured on the bright limb or edge of the moon, erupting over a caldera (volcanic depression) named Pillan Patera. The second plume, seen near the terminator, the boundary between day and night, is called Prometheus (after the Greek fire god). The shadow of the airborne plume can be seen extending to the right of the eruption vent (The vent is near the centre of the bright and dark rings). (NASA/JPL/University of Arizona) **b)** Schematic of sulphur volcano on Io (NASA/JPL/Lowell Observatory).

spacecraft that explored the moons of Jupiter between 1996 and 2003 (Figure 3.6). The earliest sightings of eruptions were witnessed by the Voyager spacecraft in 1979. Some hundred volcanoes have been detected so far on Io, and more are still to be documented. This moon is so volcanically active that its surface has been completely re-covered, leaving no impact craters such as those that scar the other planets of our solar system. The surface of Io is coloured yellow, red, orange and black, suggesting that sulphur (S) and sulphur dioxide (SO_2) must play an important role in the volcanism, an association observed at volcanoes on our own planet. Whether Io erupts pure sulphur or has sulphur-rich silicate magmas has been debated, but temperature estimates from surface eruptions are high (~700–1500 °C), supporting a silicate-based volcanism.

Earth as the home of volcanism

The Earth may not have the largest volcano in our solar system but it probably possesses the greatest variety. Due to the planet's system of plate tectonics there is a mechanism that generates a number of different volcanic environments. Additionally there is the role of hot spots/plumes, which highlight the convective cooling beneath our feet and the process of the Earth cooling. With volcanoes ranging from underwater vents to towering explosive giants and even bigger sleeping monsters, there is a volcano for almost anyone's appetite. Whether you want to marvel at one of Earth's natural firework shows, or wonder at the potential for supereruptions, the Earth is the real home for volcanoes, and a natural laboratory for us to investigate the processes and products of volcanism.

4　Types and scales of eruption

How big do eruptions get? What were the biggest ever volcanic eruptions? These are good questions, but the very size of an eruption may not be the most important factor in terms of what it can destroy and how it can change and modify the Earth's environment and its own locale. Clearly the bigger the volcano, the greater its capacity to affect the planet and the environment around it, but the relationship between eruptions and their effects is not simple. The different types of volcanoes found on Earth are classified in terms of their major morphological features and, to some extent, where they occur. The different styles of eruption from volcanoes may also be classified (some examples are

Figure 4.1 Pictures of different styles of volcano. **a)** Eruption and scoria/cinder cone on Stromboli. **b)** Fissure eruption at Fimmvörðuháls Iceland. The image is taken just a few minutes after the fissure opened on 31 March 2010 (Jón Viðar Sigurðsson). **c)** The stratovolcano Colima, Mexico.

shown in Figure 4.1). One might expect that the type of eruption and the type of volcano would be synonymous, but different types of volcano can share the same or similar styles of eruption. In the ancient rock record, where the original volcano is not preserved, the different eruption styles that formed particular deposits need to be looked at to gain an insight into what type of volcano may have formed them.

When it comes to the size of the eruption itself, the important factor in some instances is the height to which the eruption column reaches in the atmosphere. The volume of **juvenile magmatic material** that is involved in the eruptions is another way of measuring the size of eruptions. As science witnesses more and more contemporary eruptions on the Earth, our understanding of the styles and sizes of volcanic event can be revised. This, in turn, develops the classification schemes used to describe volcanoes. It is worth bearing in mind, however, that there is evidence in the Earth's rock record of eruptions much bigger than those recorded **historically**. Our understanding of the possible size of such 'super' eruptions will be constrained by the ancient volcanic deposits available for study, until eruptions of a similar scale are witnessed at first hand on the planet.

Types of volcano

From the smallest squirts to the biggest explosions, each volcano and eruption is fascinating. The different types or styles of volcano generally reflect the type of magma that they erupt and are often linked, though not always, to their position on the Earth's plates. The more viscous, or 'sticky', explosive volcanoes tend to be at destructive plate margins and the less viscous, 'runny', volcanoes are often located at constructive plate margins or hot spots. First, volcanoes can be considered in terms of their morphology, or how they look. This first step is important, as volcanoes can have very similar morphologies but have quite different scales, as demonstrated by examples ranging from the smallest basalt lava flows to some of the largest basalt flows, known as flood basalts. The basic categories of volcano are presented in Figure 4.2.

Fissure vents form as linear features on the Earth's surface and essentially represent lava erupting out of faults or cracks in the Earth's crust. Some of the most spectacular examples of fissure vents can be found at constructive plate margins where the crust is being ripped apart under extension, for example in Iceland and in the Afar Rift in Ethiopia, but they are also common in hot-spot volcanoes like those on Hawaii. Fissure-fed eruptions will often start as a sheet of magma erupting along the fault and very quickly localise to a number of more discrete vents along the fissure. Examples of well-known fissure eruptions include the Laki eruption in 1783 (Chapter 8). **Shield volcanoes** are large volcanoes formed mainly from basaltic lavas, and have a very broad relief with a very low angle to their flanks (**constructive surfaces**) due mainly to the low viscosity of the lavas that make them up. Although they have a relatively gentle profile, some examples of shield volcanoes can reach great heights due to their immense size, as with Mauna Loa, Hawaii at 4 169 m (13 679 ft.). Shield volcanoes are often associated with fissures, commonly with fissure

Volcano Type	Charactersitics	Examples	Simplified Diagram
Flood or Plateau Basalt	Very liquid lava; flows very widespread; emitted mainly from fissure eruptions	Columbia River-USA, Siberian Traps-Russia, Paraná-Etendeka - Brazil/Namibia	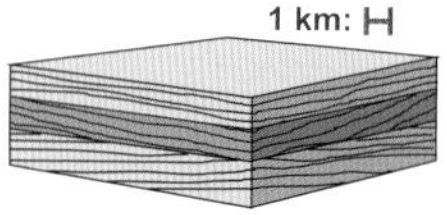 1 km: ⊢⊣
Shield Volcano	Liquid lava emitted from a central vent; large; sometimes has a collapse caldera	Hawaiian volcanoes, Erta Ale-Ethiopia, Olypus Mons-Mars	 ⊢⊣
Cinder cone	Explosive liquid lava; small; emitted from a central vent; if continued long enough may build up a shield volcano	Craters of the Moon-USA, Paricutin-Mexico, Massive central-France	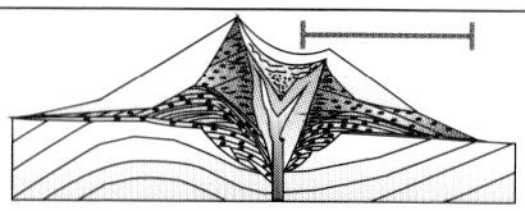 ⊢——⊣
Composite or Stratovolcano	More viscous lavas, much explosive (pyroclastic) debris; large, emitted from central vent	Mount Rainier-USA, Colima-Mexico, Damavand-Iran Mount Fuji-Japan Cotopaxi-Ecuador	 ⊢⊣
Volcanic Dome	Very viscous lava, relatively small; can be explosive; commonly occurs adjacent to/within craters of composite volcanoes	Mt StHelens dome-USA, Colima dome-Mexico, Mount Lassen-USA	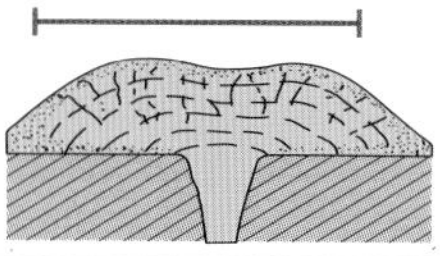 ⊢——⊣
Caldera	Very large composite volcano collapsed after an explosive period; frequently associated with resurgent domes	Crater lake-USA, Long Valley-USA Santorini-Greece	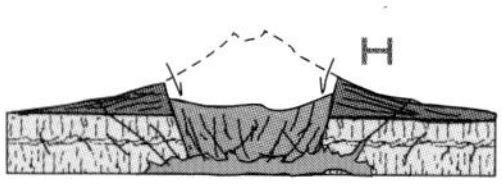 ⊢⊣

Increasing Violence / Increasing Viscosity (vertical axis label on left, pointing downward)

Figure 4.2 The basic classification of different volcanoes (adapted from the USGS).

eruptions along their flanks (Figure 4.1). **Lava domes** are bulges of new lava that form in the craters of volcanoes. These often occur after large explosive eruptions and can be termed **resurgent domes**. They tend to be relatively steep-sided, rounded mounds of predominantly viscous lava, and their collapse can lead to pyroclastic flows (*see* Chapter 5 for a more detailed description of domes). A **cryptodome** is where the volcano itself starts to inflate and bulge, usually prior to an eruption. Inflation and bulging of parts

of a volcano can cause it to become unstable and collapse, leading to the triggering of eruptions, as happened at Mt St Helens in Washington State, USA in 1980.

Cinder/scoria cones develop from small to moderate scale eruptions where pyroclastic scoria are erupted in **fire fountains** and **Strombolian** type eruptions (described in detail below). These cones are very common and can be found as isolated or clustered vents in volcanic fields or along fissures. Examples that are found on the sides of much larger volcanoes are termed **parasitic cones**. Many scoria cones are called **monogenetic** in that they only erupt once, but the cone remains. They form steep-sided mounds of scoria, splatter, **bombs** and finer material (*see* description of pyroclastic material in Chapter 6), though older cones may appear less steep due to erosion. The cones can be up to 400 m or so in height, although they are often smaller. An example of such a cone is the one formed from the 1943 to 1952 eruption of the Parícutin scoria cone in Mexico. The resultant lava flows from the cone engulfed most of the town, stopping at the church (*see* Figure 10.1b, Chapter 10), which is now a tourist stop and place of pilgrimage for those that think the church saved the town.

Composite volcanoes (Stratovolcanoes), are the big ones when it comes to volcanic landforms, and they form the towering volcanoes of children's imagination (Figure 4.1c). Rising up from shallow slopes at the base, the steep-sided tops of the volcano result in a cone-shaped mountain, often with a surprisingly small crater at the top. The sides of the volcanoes are littered with a mixture of both lava flows and pyroclastic debris. Commonly a number of valleys are observed in the sides of composite volcanoes. These channel pyroclastic flows and debris flows from the volcano. These are some of the most deadly types of volcano, and are associated with the largest types of eruptions, called **Plinian** (*see* description below). Stratovolcanoes are mainly acidic (**silicic**) in composition, with rhyolites and dacites, hence their association with explosive eruptions; but they can have the full compositional spectrum including andesites and basalts. Mt St Helens (USA),

Mt Fuji (Japan), Damavand (Iran), **Popocatepetl** (Mexico) and **Colima** (Mexico) are some examples of composite stratovolcanoes. **Calderas/caldera volcanoes** are a type of volcanic landform where the profile of the volcano, instead of being that of a classic cone or conical shape, reflects a circular/semicircular depression on the Earth's surface. Such depressions, or calderas, are formed by the collapse of the volcano after the evacuation of magma from a shallow chamber that was present before the eruption. Once the magma has been erupted, the void left by the removal of the magma is filled by the collapsing carapace above. Calderas can be associated with many volcanic settings, but tend to occur with very large eruptions (see 'Giant holes in the ground' section later).

Submarine volcanoes occur when a volcano starts to erupt on the sea floor, and builds up as a seamount. In some instances the growth is such that the volcano breaks through the ocean surface with dramatic effects. These spectacular eruptions are where land is made for the first time from the sea. The submerged island **Ferdinandea** (Ìsula Firdinandèa) 30 km south of Sicily is an example of one of these submarine volcanoes that is fluctuating above and below sea level, a so-called **'pop-up'** volcano. It last rose above sea level in 1831, leading to a dispute over its sovereignty, which was somewhat short-lived, as the waves it initially broke through levelled it off, and the island disappeared beneath the waves by 1832. In 2010, the Princess of Sicily took part in a ceremony to place a plaque on the shallow summit of the volcano to claim the future land for Sicily. Other examples have occurred recently – off the coast of

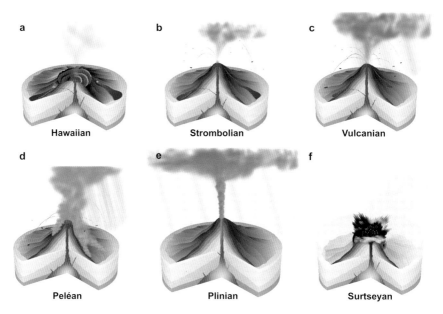

Figure 4.3 Key types of volcanic eruption. **a)** Hawaiian; **b)** Strombolian; **c)** Vulcanian; **d)** Peléan; **e)** Plinian; **f)** Surtseyan (3D images by Ivan d'Hostingue, aka Sémhur, inspired by the document about volcanism from the web portal on the prevention of the major risks by the French Minister of Ecology, Environment and Sustainable Development).

Tonga in the Pacific, and the formation of the Island of Surtsey off Iceland (*see* Chapter 6).

Subglacial volcanoes, where an eruption occurs under ice, are very complex in that they involve interactions between magma, water and ice. They are also constrained by the ice that covers them. Examples around the world are restricted to areas where volcanism interacts with ice now, or where it interacted with ice sheets in the past. Mainly studied in Iceland, these volcanoes, when exposed after ice retreat, have a flat table-topped shape and are known as '**Tuyas**' from the mountain in British Columbia, Canada. Modern examples produce spectacular eruptions that can lead to massive flood events

known as **Jökulhlaups** as the volcano melts the ice water, and can produce high eruption columns. The famous 2010 eruption of Eyjaf-jallajökull in Iceland disrupted European airspace (*see* Chapter 10). **Mud volcanoes**, named after volcanoes but not involving magma, are the result of hot over-pressured mud and groundwaters that can erupt to produce mini volcanic-like landforms, with excellent examples found in Azerbaijan (*see* Chapter 5).

Types and scales of volcanic eruption

It may not be the size that matters, but people are always fascinated by how big volcanic eruptions can be. It has been noted for many years that different types of volcano can result in different sizes of eruption, with the more silicic 'sticky' volcanoes tending to have the more explosive activity. A general classification of eruptions has developed, based on and named after key volcanic events, which captures the general scales of eruption based on observations of the eruption column and the explosiveness of the event (Figures 4.3, 4.4). There is also a more detailed classification known as the Volcanic Explosivity Index (described later). Types of eruption range from those where the eruptions are not very explosive and have little or no eruption column to the largest type of explosive activity. An additional factor that can affect the style of eruption is the interaction with water or ice. So as the type of volcano can be defined, also the different types of volcanic eruption can be classified.

Effusive eruptions are the very simplest of volcanic eruption where lava flows out of a vent, forming a lava flow, and no explosive activity is involved. Such eruptions can occur commonly at **lava lakes** when they fill up and spill over, but it is rare for a predominantly effusive eruption to develop without some sort of more explosive event at its start. **Hawaiian eruptions** are the more common type of activity that is seen in basalt-dominated volcanoes. They are intimately associated with effusive eruptions, which is often the style of eruption that can feed much larger lava flows through time. Lava fountains that reach a few hundred metres into the air are common, and examples of this style of eruption along fissures can sometimes result in a 'wall of fire' which collapses to the ground to feed lava flows. Classic examples of this type of eruption are the

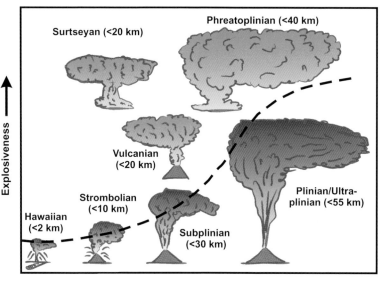

Figure 4.4 Height versus explosivity for key types of eruption. Degree of explosivity/fragmentation plotted against the observed eruption column height (which can be estimated for unobserved/ancient deposits). The addition of water aids the fragmentation of the juvenile magma.

Kīlauea Iki eruption of 1959 in Hawaii, and the Krafla 1975–83 eruption in Iceland. **Strombolian eruptions**, some would say, are the most picturesque of volcanic styles, named after the **volcano** forming the Island of Stromboli, in the Mediterranean. A Strombolian eruption is driven by the bursting of gas bubbles within the magma. These gas bubbles accumulate as they ascend in the volcano conduit and coalesce into larger bubbles, called 'gas slugs', which burst at the surface. This can result in a variety of explosions, from small volcanic blasts to larger eruptive columns. Strombolian activity is characterised by scoria and bombs of hot material being ejected as the bubble bursts. In extreme cases, columns can measure hundreds of metres to kilometres in height but, more commonly, a Strombolian eruption looks somewhat like an energetic Roman candle firework when viewed at night (Figure 4.1).

Vulcanian eruptions, named after the 1888–1890 eruptions on Vulcano Island in the Aeolian Islands north of Sicily in the Mediterranean, are more explosive still. These eruptions result from more viscous magma (andesite/dacite), where it is difficult for gases to escape easily. The pressure builds up and eventually the blocked top of the volcano gives way with an initial set of cannon-like explosions, sending bombs and blocks through the air. The trapped gas in the underlying magma expands dramatically as it reaches the surface and fragments the magma, resulting in a large amount of ash. The deposits contain a much larger amount of ash than their Strombolian counterparts, with eruptive columns reaching up to ten kilometres. Some of the high gas contents in

Vulcanian eruptions have been attributed to the interaction of the magma with groundwater, known as **hydrovolcanic eruptions**.

Plinian eruptions are the most spectacular of the eruption styles that are seen on Earth. They are named from the famous AD 79 eruption of Mt Vesuvius, observed by Pliny the Elder and **Pliny the younger**. These eruptions are very powerful and result in a large column of ash and **ejecta** and a cloud of ash that travels high up into the atmosphere (up to 55 km) and spreads out where it reaches the stratosphere, to form a mushroom-like shape that was likened to an umbrella pine tree by Pliny when he described the cloud at Vesuvius (*see* Chapter 10). This was the first description of this type of eruption, and the volcanic cloud type is known as a **Plinian cloud** from these original descriptions. The column of ash and ejecta in a Plinian eruption can periodically collapse and lead to pyroclastic eruptions, which can travel at high speeds. These eruptions form from violent volcanic phases in stratovolcanoes, and can also be associated with partial or complete destruction of the volcano itself.

Peléan eruptions, which are named after the 1902 explosive eruption of **Mt Pelée** in Martinique, are similar in scale to Vulcanian eruptions but are characterised by hot glowing clouds (termed **nuée ardente**, meaning glowing cloud) of pyroclastic flows. These may be driven by the collapse of domes that form in the volcano's crater, which over-steepen and collapse, or by the collapse of Plinian columns. The 1902 eruption was preceded by a thick lava spine that bulged out of the volcano. When it collapsed the resultant pyroclastic flows engulfed the town of St Pierre, and killed more than 30 000 people in one of the worst volcanic events in the twentieth century.

Surtseyan eruptions describe the type of explosive volcanism, often forming a **tuff cone**, that occurs when a submarine volcano erupts through the waves to produce new volcanic islands. This is named after the example that occurred off the coast of Iceland in 1963, leading to the formation of the Island of Surtsey (also Chapter 6). In this example the eruption was very explosive to start with, but as the island started to build, the eruption became more Strombolian in its nature. More generally they are known as **hydro-volcanic** (**phreatic** or **phreatomagmatic**) eruptions, which signifies the interaction of lava with water.

Two factors that are important to consider when looking into the overall effects of volcanic eruptions are their explosiveness and the **eruption column**. The eruption column is easy to measure in observed eruptions and can be estimated in non-observed and ancient deposits by looking at the area of the dispersed pyroclastic material. The explosivity is a measure of how fragmented the volcanic tephra (ash) is from the eruption (measured as the amount (%) of particles < 1 mm at a point from the volcano where the deposit thickness is 10% of the maximum thickness reached). This provides a good graphical way to explore the different styles of eruption and gives an idea of their magnitude.

Table 4.1 Volcanic Explosivity Index, VEI. Classification table of the size of volcanic eruptions using the VEI. (Smithsonian Institute database)

VEI	Ejecta volume	Classification	Description	Plume	Frequency	Some Examples	Occurrences in last 10,000 years[*]
0	< 10,000 m³	Hawaiian	non-explosive	< 100 m	constant	Kilauea	many
1	> 10,000 m³	Hawaiian/Strombolian	gentle	100–1000 m	daily	Stromboli	many
2	> 1,000,000 m³	Strombolian/Vulcanian	explosive	1–5 km	weekly	Tristan da Cunha (1961) Whakaari/White Is. (2019)	< 4250
3	> 10,000,000 m³	Vulcanian/Peléan	severe	3–15 km	yearly	Surtsey (1963) Soufrière Hills (1995) Ontake (2014) Anak Krakatoa (2018)	989
4	> 0.1 km³	Peléan/Plinian	cataclysmic	10–25 km	≥ 10 yrs	Mount Pelée (1902) Eyjafjallajökull (2010) Nabro (2011)	461
5	> 1 km³	Plinian	paroxysmal	> 25 km	≥ 50 yrs	Mount St. Helens (1980) Mount Hudson (1991)	174
6	> 10 km³	Plinian/Ultra-Plinian	colossal	> 25 km	≥ 100 yrs	Krakatoa (1883) Mount Pinatubo (1991)	52
7	> 100 km³	Plinian/Ultra-Plinian	super-colossal	> 25 km	≥ 1000 yrs	Tambora (1815)	5 (+2 suspected)
8	> 1,000 km³	Ultra-Plinian	mega-colossal	> 25 km	≥ 10,000 yrs	Toba (71–72,000 BP)	0

[*]Count of VEI 2 and VEI 3 eruptions in the last 10,000 years are based on summary 1994 figures maintained by the Global Volcanism Program of the Smithsonian Institution, with additions from the new 1995–2000 database. Count of eruptions greater than VEI 3 in the last 10,000 years are based on its 2010 figures, with additions based from the 2020 database update. There are also 58 plinian eruptions, and 13 caldera-forming eruptions, of large, but unknown magnitudes. Note: There is a discontinuity in the definition of the VEI between indices 1 and 2. The lower border of the volume of ejecta jumps by a factor of 100 for VEI between all higher indices (Information sourced from Smithsonian Institute, USGS).

Interestingly, the hydrovolcanic eruptions that plot above the dashed line (Figure 4.4) can produce highly fragmented examples, but do not always produce the highest eruption columns.

In order to indicate some measure of the relative sizes of explosive volcanic eruptions, a classification scheme called the **Volcanic Explosivity Index (VEI)** has been developed. This uses a number of criteria to estimate the overall force of a volcanic event. The VEI runs from 0 where eruption is not explosive through small (1), moderate (2), moderate–large (3), large (4) and very large (5–8). These different sizes in turn are related to a type or style of eruption (Table 4.1). As the table shows, the different types of eruption discussed above can be placed into the VEI with relative descriptions of their explosivity and the height of the eruption column (as in Figure 4.4). Additionally, some attempt has been made to work out the relative frequency of such events, with many examples of eruptions with low VEI and very few examples with a high VEI. In detail, the VEI is calculated through a large range of data that are collected for individual eruptions through careful observation and mapping out of the erupted products. Primarily these include the volume of erupted material, the height of the column and the duration of the eruption. It is even possible to calculate the VEI of ancient deposits if they are mapped out well enough, as one can correlate the size of **clasts** and ash with distance to estimate column height for unobserved eruptions. A graphical way of thinking about the VEI is given in Figure 4.5. The Klyuchevskoy volcano in Kamchatka, as so beautifully imaged from space (Figure 4.6), had a Volcanic Explosivity Index (VEI) of 3, Lava Volume of $3.0 \pm 1.0 \times 10^7$ m^3,

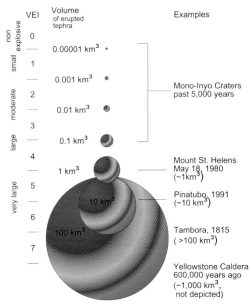

Figure 4.5 Volcanic explosivity index (VEI) in graphical form. The relative size of the sphere relates to the volume of the eruption example given for specific VEI (USGS).

Figure 4.6 Eruption column of volcano viewed from space – the October 1994 eruption of Klyuchevskoy Volcano in northern Kamchatka, Russia. This explosive eruption sent a column of volcanic ash more than 15 km (9 miles) into the atmosphere. This photograph is taken from a space shuttle showing the ash cloud being blown south-eastward over the North Pacific Ocean (NASA).

Tephra Volume of 5 x 107 m³. The recent Iceland volcanic eruption had a VEI of 4 with an estimated 250 million m³ (¼ km³) of ejected tephra and an ash plume of ~ 9 km (30 000 ft).

The timing of eruptions

Frequency and timing of eruption is important, as is how often a volcano or range of volcanoes will be active or dormant. Chapter 9 looks in some detail into the way in which the onset of eruptions is monitored. Here the consideration is how often do volcanoes erupt, and are they cyclic in their behaviour? During the active eruption of a single volcano, it may remain quiet for relatively short periods of time, minutes, hours or days, and then explode into life repeatedly, as if to suggest some cyclicity within a single

eruption phase. On a longer timescale a single volcano may erupt several times between much larger gaps, years to thousands of years, during which time it may be considered dormant (or even, possibly incorrectly, extinct). How geographically close volcanoes may or may not be linked is another question, the answer to which relies on our being able to date an eruption accurately. The possible data that can be used to get relative and absolute timing of events that occur either within a single eruptive episode at a volcano or in piecing together its past, and the past activity of volcanoes around it, is wide and varied. There are observations made directly as a result of eyewitness accounts and the monitoring of eruptions that happen today or which happened in the very recent past, but

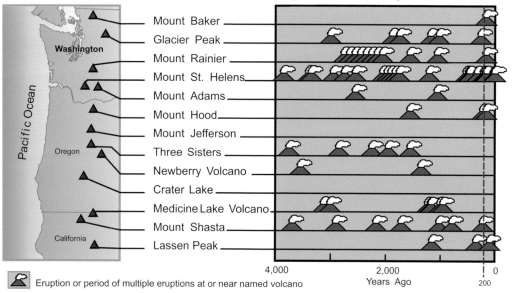

Eruptions in the Cascade range During the Past 4,000 years

Figure 4.7 Timing of eruptions from the Cascades range, USA. Compilation of eruptions in the Cascades range over the last 4000 years (USGS).

these get more sketchy the further back in time they occurred. Where there are eruptive products, such as lava flows and pyroclastic rocks, and where there is no eyewitness information, there are other methods of dating these eruptions, such as **carbon dating**, where any wood that was caught up in the eruption can be dated using its carbon isotopes. From observation and monitoring of eruptions as they happen, some clear short timescale cyclicity can be observed. Stromboli, for example, erupts every 15–20 minutes or so, and is thought to be regularly degassing with intermittent slugs of gas that accumulate and then release at fairly regular time intervals. Similar patterns of degassing and cyclic activity are seen at Colima in Mexico, which erupts gas and ash clouds at regular intervals. **Soufrière Hills** Volcano, **Montserrat**, has been shown to have cycles of dome growth and degassing in the order of 11–16 days as parts of larger phases of dome growth and collapse that have marked the recent eruptive history of this volcano between 2005 and 2007. These are examples where the regularity, or not, of the eruptions may be observed to see how they develop (Chapter 9), but looking at the longer timescale at a single volcano and at its surrounding neighbours is a much more difficult prospect.

One of the areas where good data have been compiled about the longer-scale timings of volcanic products from a number of closely spaced volcanoes is the **Cascades** range in the USA. This has seen one of the most widely studied eruptions, that of **Mt St Helens** in 1980, but the backbone of work has been carried out by the United States Geological Survey (USGS), universities and other organisations, looking at the historic and preserved record from several different volcanoes. Figure 4.7 presents the results from this compilation by USGS and provides a good visual record of the distribution and relative timings of the activity of the Cascades volcanoes. Some of the volcanoes, for example Mt St Helens and Mt Rainier, erupt very regularly. Others have long breaks between eruptions; Mt Adams, Newberry Volcano; whilst some may be showing clear cyclic patterns; Mt St Helens, Mt Shasta and Glacier Peak. It is important to realise that the compilation of such data-sets takes a lot of time and effort with not only well documented eruptions, but also those pieced together from the recent geological past. They rely on accurate dating techniques to constrain the eruptive past of the volcanoes. As systems are studied, the relative timings and potential cyclicity of volcanic systems over a variety of timescales can be pieced together, with the intention of being better able to form future predictions. Linking two or more volcanoes together in terms of eruptive events that seem to coincide is potentially trickier still; *see* Chapter 10 when we look at Eyjafjallajökull on Iceland and its possible links with the neighbouring volcano Katla.

Giant holes in the ground

You will have seen the term 'caldera', as defined in Figure 4.2 and earlier within the text; but what exactly are they, and how are they formed? This at first seems quite straightforward, as the escape of magma during a big eruption would leave a void space where the ground can collapse into. Indeed this is thought to be how some of the biggest calderas are created in predominantly silicic systems. They can form giant holes in the ground that can sometimes only be appreciated from space, such as those in Campi Flegrei (Italy) and Yellowstone (USA).

One of the classic examples of a caldera is that of the famous Crater Lake in Oregon, USA (Figure 4.8). Here an ancient mountain, known as 'Mount Mazama', was destroyed in an explosive volcanic event some 7700 years ago, which removed the top from the estimated ~3350 m (12 000 ft) high Mount Mazama, resulting in the fantastic Crater Lake Caldera we see today. This is not only evidenced by the Caldera, but also by the vast pyroclastic deposits (*see* Chapter 6) that are deposited around it. However, some of the most recent 'caldera' forming eruptions are somewhat more surprising in the way they have formed. Examples seen within predominantly basaltic systems can show a strange separation between the place of subsidence (the caldera), and the site of eruption – none more spectacular than that of the recent Icelandic **Holuhraun** eruption and its associated caldera collapse at **Bárðarbunga** volcano. Here magma travelled away from the sub-volcanic magma system and channelled

Figure 4.8 The formation of Crater Lake Caldera. a) Schematic stages of the eruption sequences that resulted in the evacuation of the magma chamber beneath Mount Mazama, and the formation of the caldera. **b)** View across Crater Lake with Wizard Island in the foreground. **c)** Digital perspective view of the generalised geologic map of the lake floor draped over shaded relief image of 2-m bathymetry, revealing the extent of the caldera (USGS).

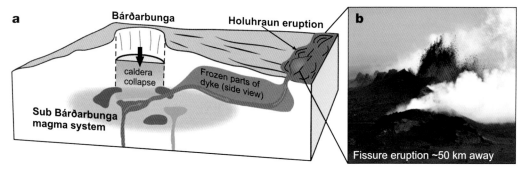

a Bárðarbunga Holuhraun eruption **b**

caldera collapse

Frozen parts of dyke (side view)

Sub Bárðarbunga magma system

Fissure eruption ~50 km away

Figure 4.9 Caldera collapse through flank eruption at Bárðarbunga volcano in Iceland. **a)** Schematic 3D view of the plumbing system beneath the volcano highlighting the magma pathway away from the volcanic centre to erupt on the flanks some 50 km away (3D template — Freysteinn Sigmundsson). **b)** The fire-fountaining Holuhraun fissure eruption, pictured in September 2014.

underground along a sideways dyke system before eruption at the surface nearly 50 km away from the volcano, as a spectacular fire-fountain fed lava flow (Figure 4.9; *see* also Chapter 10 'Birth of a Volcano'). Hence the volcano collapsed to form a caldera, yet the eruption occurred along the flank of the system, an association that may be a more common feature than we have fully understood, and one which may be typical of large volcanoes associated with rift systems like Iceland and Hawaii.

The largest eruptions on Earth – LIPS and supervolcanoes

The largest eruptions to have occurred on Earth are not necessarily the most explosive ones, but are commonly made up of basalt rocks similar to those that erupt on Iceland and Hawaii today. These volcanic events are known as flood basalts and have occurred at key points in the Earth's history. Some of the largest individual eruptions are measured in thousands and tens of thousands of cubic kilometres and covered vast areas of land. Through **geological time** the Earth has

experienced a number of large outpourings of lavas, giving rise to **Large Igneous Provinces (LIPs)** which have been mapped out to show their distribution (Figure 4.10). These LIPs or flood basalt provinces (a term used when they occur on land) have been associated with significant events in Earth's geological history such as the breakup of **supercontinents** like **Gondwana**. They have been linked also with **extinction events**. Chapter 8 looks in more detail at how these large eruptions and volcanoes in general can change the environment. These large LIP events have been linked with hot plumes of rising mantle to generate such large volumes of melt, with some provinces exceeding one million cubic kilometres of magma. The term '**supervolcano**' has been used to describe a volcano that has the capability of erupting more than a thousand cubic kilometres of ejecta in a single event (100–1000 times larger than historic eruptions). Clearly the eruptions associated with LIPs fit into this category, and six current volcanoes have been identified as possible supervolcanoes, based on their previous activity: Yellowstone,

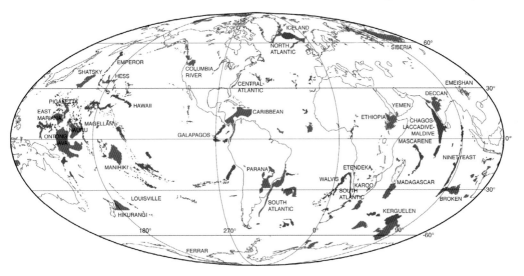

Figure 4.10 Large igneous provinces (LIPs) of the world map. Phanerozoic global LIP distribution, with transient main pulse ('plume head') in red and persistent ('plume tail') in blue (Mike Coffin, 2006).

Long Valley, and Valles Caldera in the USA; Lake Toba, Indonesia; Taupo Volcano, New Zealand; and Aira Caldera, Japan. The largest eruptions would have explosive indices at VEI-8. There has been much media speculation trying to ascertain what would happen if such an eruption happened today. The effects of an explosive VEI-8 would clearly be different to an effusive lava flow eruption of similar volume, but it is hard to know which would have more long-term impact on the planet. What is clear is that the short-term effects on the environments close to such eruptions would be devastating. Large volume eruptions and climate are the subject of Chapter 8.

5 Lava flows and bubbling cauldrons

Molten rock erupting at the surface can either be exploded into fragments, giving rise to pyroclastic rocks, or may be emitted more passively on the surface as lava flows. In truth there is no simple divide, as in some cases lava can flow from fire-fountain explosive type eruptions when they fall back to the ground, providing the magma does not cool sufficiently before re-forming as molten lava.

Lava can erupt and flow under water. It can boil over from long-lived lava lakes or drain from faults in the sides of mountains. This chapter looks at all the aspects of lava, from the bubbling cauldrons that are lava lakes to the vast flows that can cover vast areas of the planet and the stages in between (Figure 5.1).

Just as lavas can form from explosive pyroclastic eruptions, so, too, can volcaniclastic

Figure 5.1 Spectacular pictures of lava flows: **a)** Fresh pāhoehoe lava skin forming, Hawaii; **b)** Gas bubble bursting on the surface of Erta Ale lava lake, Ethiopia (photo Rupert Smith); **c)** Lava 'break-out' on pāhoehoe lava field, Hawaii; **d)** Lava flows from the Fimmvörðuháls eruption falling down into Hruná-canyon, Iceland, on 31 March 2010 (photo by Jón Viðar Sigurðsson).

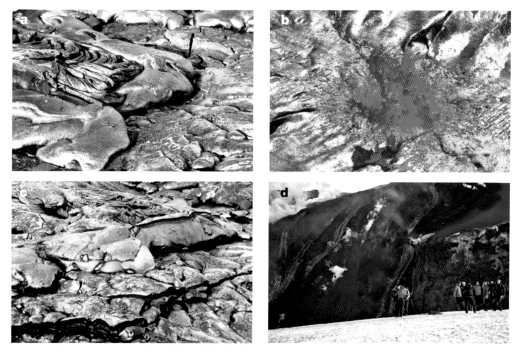

rock develop from lava flows if they fragment due to changes in viscosity or, in some examples, when interacting with water or ice. On the whole, volcanic rocks can be separated into those formed from lava flows and those formed by explosive volcanic activity, which are dealt with in the next chapter. 'How' lavas work, how the different compositions of magma can affect the style of lava flow, and how their interaction with water can change eruption and flow properties needs to be understood. The odd aspects of lavas on the surface of the Earth will be examined from the ancient lava lakes that occur on our planet to eruptions with odd compositions, from carbonate-rich volcanoes to those that erupt mud.

How do lava flows work?

It is tempting to think of lava flowing like water in rivers, as is often depicted in the Hollywood blockbusters, but lava flowing like rivers in open channels is not at all common. Clearly, when lava is flowing fast down a steep slope, or where it is in a very constrained channel, it will behave like any other liquid, but more commonly the cooling crust of the lava plays a very important role in how the lava flow develops. A basaltic lava will be at a temperature of 1100 + °C when molten. Unless it is moving very quickly, the lava will soon form a cooling crust on top. This is one of the key factors that enable lava flows to travel a long way and to attain thickness. From observations of lava flows on Hawaii, it was demonstrated that a mechanism of **lava flow inflation** was the key to understanding how sheets of lava can develop and how most lava flows travel from their source to their end. The crust that develops on the lava insulates the molten magma beneath. As the lava flow moves forward the cooling crust thickens and soon the lava flow thickens as hot, insulated magma is pumped beneath the crust. This is also the way in which fresh, hot lava reaches the front of the lava flow and breaks out further. Once its crust has developed sufficiently, lava is thus able to travel long

Figure 5.2
Emplacement of lava flows by inflation. Shortly after the lava breaks out, a crust forms insulating the lava within. The lava grows by injection of new magma which inflates the crust of the lava, a process that can go on for days/months/years. The final lava flow has an internal structure of a massive core and a fractured crust rich in vesicles (preserved bubbles) (Adapted from Self et al., 1997).

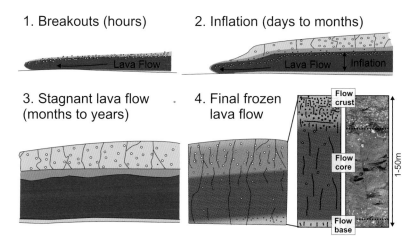

1. Breakouts (hours)

2. Inflation (days to months)

Lava Flow

Lava Flow   ↓ Inflation

3. Stagnant lava flow (months to years)

4. Final frozen lava flow

Flow crust

Flow core

Flow base

1–50m

distances without much cooling. This inflation process of lava flow formation is depicted
in Figure 5.2. This process, if continually fed
with new magma, can lead to a great thickness and a great lateral extent of flow. The
Great Greenstone Flow in Isle Royale, Michi-
gan is over 80 km and has thicknesses up to
250 m whilst flood basalts carry thousands of
cubic kilometres of lava hundreds of kilometres from their source with little obvious
cooling.

The runny ones: basic flows

Basalt is by far the most common form of lava
on the Earth's surface. Under abyssal sediment it forms the top parts of the ocean crust
and is found in almost all volcanic environments. The composition of basalt, with relatively low silica contents (45–55% SiO_2), gives
it a relatively low **viscosity**, so it can be considered as a 'runny' lava, and can cover areas
as large inflated sheet flows or form 'rivers' of
lava like giant braided river channels (Figure
5.3a). This is clearly a generalisation, but as
the silica content decreases, a decrease in viscosity is observed. In the main, basaltic flows
behave like low viscosity magmas. The most
common types of lava flow observed on land
form sheets and overlapping lobes with a
characteristic 'rope-like' texture on their top
surface known as 'pahoehoe' (Figure 5.3b).
The term pahoehoe is Hawaiian, meaning
'smooth, unbroken lava', and this type of
flow is very typical of the inflation mechanism. The ropy texture is formed as a very
thin crust develops and then is rippled by the
hot lava flowing beneath. The flows move
forward by a series of '**breakouts**' where hot
lava breaks through the crust and starts to
flow again before it cools significantly and

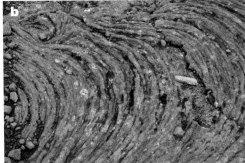

Figure 5.3 Examples of basic flows. **a)** Braided river-like
channels of basaltic lava from the 2018 Hawaiian eruption
(USGS). **b)** Ropey pāhoehoe lava surface, Iceland. **c)**
Measuring the thickness of an 'a'a flow produced by the
Kamoamoa fissure eruption, Hawaii. The measuring stick
is 2 m (6.5 ft) tall (USGS).

forms new crust. This process is repeated again and again, resulting in the intricate overlapping lobes on the top of pahoehoe flows (Figure 5.1a & c). The tops of the lava sheets are also often riddled with cracked dome-like features known as '**tumuli**'. Lava flow tumuli are small 'mini-volcanoes' that have popped out along the tops of the lava sheets as more significant breakouts from over-pressured lava blisters. These form pock marks, like spots, on the generally smoother lava surface.

Another common lava type found in basalts is known as **a'a** (meaning 'stony rough lava' also from Hawaiian). In this case the lava surface, instead of being smooth, is characterised by irregular blocks and sharp fragments of lava (Figure 5.3c). These spiny, broken pieces of lava resemble the clinker from a smelting slag heap in a jumbled collection of irregular, loose but clearly fresh material, as if spalled or ripped from the lava.

The front of an 'a'a' flow advances like a big conveyer belt, with the brecciated lava fragments tumbling down the face of the advancing lava, as if being carried to the front of the flow by more cohesive lava. Examples of a'a lavas are shown in Figure 5.3. Both pahoehoe and a'a lavas are found in the same lava flow fields, as well as transitional types such as 'blocky pahoehoe'. There is a complete spectrum between both types, with the transition from pahoehoe to a'a marking a change in viscosity that is thought to be caused by the crystallisation of very small crystals, **microlites**, which can have a dramatic effect on the viscosity of the lava, making it more susceptible to tearing. Such a viscosity change can also be caused by the arrival of new lavas with a more silica-rich composition.

The sticky ones: acidic flows

When the silica composition of a magma is increased, then it becomes much more viscous

Figure 5.4 Lava flow features from Colima, Mexico. Lava flow and drainage levees down the side of Colima Volcano, Mexico.

levees

Lava flow

or 'sticky', and will flow more slowly. In addition to this, the higher the viscosity the harder it is for gases to escape from the magma, so often the volcanoes that have higher silica content (acidic or **felsic** are commonly used terms), are characterised by explosive eruptive styles. Where lava flows out from such volcanoes it is somewhat different to that in the more common 'basic' eruptions discussed already. The lava flows formed from acidic compositions tend to have very high aspect ratios and rarely move significant distances (a few kilometres) from the vent area due to their viscosity and, usually, relatively low eruption volume. The edges of the flows can be several tens to hundreds of metres thick, and on steep sides the flows can form levées where the central parts of the flow have continued to flow whilst the edges have cooled (Figure 5.4). The flows are made up of cuspate blocks of lava. Inside, complicated ramps and 'buckled up' parts of the lava occur as a complex pattern of **fractures** and **joints** develop with the different strains of cooling and the advancing viscous flow. The lava itself tends to be very glassy and in many cases, the whole of the lava forms a natural glass known as **obsidian**. Classic obsidian-rich flows include those on Lipari Island in the Aeolian Islands and Obsidian Dome, Mammoth Lakes, USA.

Lava domes are formed in the craters of many volcanoes that erupt high-silica lavas. These are steep-sided, predominantly rounded mounds that represent the oozing out of relatively degassed magma in the central crater region (Figure 5.4). The term 'resurgent dome' has been used to signify the relationship of dome growth after a destructive volcanic episode. A resurgent dome filled Mt St Helens' crater after the 1980 eruption blew the heart out of the mountain (Figure 5.4).

Columns, bubbles and tubes: inside the flows

The internal structures of lava flows can be dramatic, and are fascinating as they demonstrate a complex relationship between the original hot flow and associated cooling-related structures. Ancient deposits provide an opportunity to see lava flows in cross section, although some examples of relatively recent flows may be exposed if there has been rapid local erosion or collapses revealing their internal structures. In some instances tubes may be followed through lava flows that have drained and cooled. These examples also allow us to see lavas in section. The internal features of lava flows may be grouped into both gross scale features, those that relate to the overall structure of the flow itself, and also to textural features on a somewhat finer scale. Some features are synonymous with a range of lava flow types, and some are more specific to a particular style of lava flow.

The gross scale features displayed by lava flows are generally dictated by the jointing patterns caused by the flow's emplacement and cooling, as well as the distribution of **vesicles** (preserved bubbles). The general structure of an inflated basalt flow (as indicated in Figure 5.2) shows that the core and crust parts of the unit differ in degree of fracturing and the amount of vesicles. This results in the crust of the flow having very different rock properties from the, essentially degassed, inflated core. Inspection of the upper parts of these flows can reveal vesicles in layers in the crust and **gas escape structures**. The most striking internal jointing

Figure 5.5 Columnar cooling joints, entablature and colonnade. **a)** Simplified entablature and colonnade columnar jointed structure found in some lava flows. **b)** Classic columnar and entablature at Fingal's Cave, Scotland. **c)** Close-up view into Fingal's Cave. **d)** Examples of these types of columnar joint structures from Iceland.

pattern seen in lava flows is that of **columnar jointing**, where cooling joints within the lava, or in some instances in shallow intrusions, result in predominantly hexagonally arranged columns (Figure 5.5). These are formed at right angles to the cooling front. Within each

Figure 5.6 Internal structures, crystals and vesicles. **a)** Filled vesicles (amygdales) from Skye, Scotland (note larger vesicle fill has a horizontal early fill indicating way-up). **b)** Layers of vesicles, 'vesicle stratigraphy', from Talisker Bay, Isle of Skye, Scotland. **c)** Large plagioclase feldspar crystals, 'phenocrysts', in a lava from the Faroe Islands. **d)** Large tabular feldspar phenocrysts in a lava from Teide Volcano, Tenerife.

column the faces maintain the minimal surfaces required to fit together in three dimensions with faces at roughly 60 degrees to each other. The formation of columns is particularly enhanced by water and, as thick flows can take many days, months or even years to solidify, eruptions that occur in wet environments display good examples of columnar jointing. Where water cooling has played a significant role, often when lava flows are 'ponded' in river valleys and are cooled by river water flowing over them, a predominantly two-tiered set of columns can develop, known as entablature and colonnade (Figure 5.5). The **colonnade** columns rise straight up

from the basal cooling form of the lava, whereas the ingress of water in the upper parts of the flow sets up a variety of different angles of cooling fronts. This leads to an irregular and sometimes hackly jointing called '**entablature**' in the upper parts of the flow, sometimes with fanning patterns around partially regularly spaced cracks that descend from the lava flow surface, enhanced by water ingress.

Textural features of lava flows may include the presence of large crystals within the flow (known as **phenocrysts**), and the presence of vesicles. Phenocrysts in particular flows can be useful as markers in monotonous lavas of

Figure 5.7 Amethyst geodes. Giant amethyst geode from geode mines in southern Brazil.

the same composition where crystals are absent. The crystals themselves are particularly important, as they can be used to establish information about the origin of the magmas in the system (*see* Chapter 2). The crystal contents of lavas can increase the viscosity of the magma, whilst the patterns and alignments of the crystals can provide localised flow patterns (Figure 5.6). The vesicles also tell an important story about the development of the lava flows. In sections through lavas repeated layers of vesicles (Figure 5.6) may be observed. Termed **vesicle stratigraphy**, these are formed by successive pressure pulses within the magma relating to the inflation mechanism. As a breakout occurs at the front of an advancing lava flow, the internal pressure within the flow drops, leading to a **vesiculation event**. The resultant bubbles rise to the top of the magma and are trapped by the hardening upper crust of the flow. As the lava flow develops further, repetitions of this process result in layers of vesicles. After burial, the vesicles within lava

flows are often filled with **secondary minerals** like quartz, **calcites** and **zeolites** which precipitate from fluids that pass through the lava; the latter can be indicative of the depth of burial. Filled vesicles in magma are termed **amygdales**, for example amygdaloidal basalt (Figure 5.6). The most spectacular filled hollows in magma are found in megavesicles in the upper parts of particular flows, for example in Brazil, where they form massive **geodes** containing pretty minerals such as **amethyst** and **agate** (Figure 5.7).

Lava flows and fields are fed by an internal feeder system referred to as **lava pathways**, forming an intricately arranged set of pipe-like feeders. When such pipes are only partly filled with magma, or when they drain due to the internal lava pouring out, a **lava tube** is the result (Figure 5.8 a & b). These tubes can be many kilometres long. They are a caver's delight, as they can contain **stalagmites**, **stalactites** and columns of lava as well as marker horizons on the walls of the tube, marking previous lava levels. Many of the lava tubes around volcanoes are now tourist attractions in their own right, and the more extreme ones pose caving challenges. The tubes can also play an important role as part of aquifer systems where volcanic rocks hold the majority of groundwater (e.g. the Snake River Plain, USA). In rare examples, the whole of a shallow magma body can drain, as seen at Thrihnukagigur in Iceland, known as the 'Inside a Volcano' experience (Figure 5.8c). Caverns in volcanically driven hydrothermal systems can also develop, with the most spectacular being the Cave of Crystals in Mexico, where giant 13 m long gypsum crystals have grown under special heated ground water conditions (Figure 5.8d).

Figure 5.8 Examples of lava tubes, caverns and volcanically driven holes. **a)** Lava tubes in the Galápagos, Santa Cruz Island; **b)** Ice sculptures in lava tube caves, Wudalianchi volcanic field, northeastern China; **c)** Giant lava cavern/drained chamber known as the 'Inside a Volcano' experience, Iceland (Morgan Jones); **d)** Author and a team from Channel 9 News in the Cave of Crystals, Mexico (Andy Taylor).

Pillows and fragments: lava into water

When lava flows into water, or indeed erupts under it, it forms distinctive deposits that are seen on the Earth's surface today and preserved in the rock record. Lava can be over 1000 °C and when it comes into contact with water it is dramatically quenched. Simultaneously, the water itself is converted into steam. There are two main examples of lava interacting with water: when a lava flow enters a body of water such as the sea or a lake, or when a submerged volcano erupts under water and builds up to the point where it can emerge to form a new island. Other interactions with water occur where volcanoes interact with ice, which can happen at volcanoes topped by glaciers, or where a significant ice sheet covers them, as occurs in Iceland. The lava itself can then fragment into breccias and finer material, or may flow through the water, as a function of the gases in the magma, the type of water interaction,

Figure 5.9 Pillow lavas and hyaloclastites. **a)** Underwater active shot of pillows forming taken from ROV (remotely operated underwater vehicle): Jason II, Lau Basin, south-west Pacific, 2009 (Marine Geoscience Data System). **b)** Cross section through pillows showing glassy rind around pillow in red/brown and internal jointing, Iceland. **c)** Lava entering the sea where it forms hyaloclastites offshore, Hawaii. (USGS) **d)** Hyaloclastite deposits, Iceland.

the slope down which the lava is flowing and the water depth.

Lava that flows into water or erupts under it without significant fragmentation results in a very distinctive type of flow known as **pillow lava**. This is a common feature on the Earth, as it makes up the top surface of the ocean crust but is only rarely found in the rock record on land, as tectonic processes destroy it (*see* Chapter 3). Pillow lavas are similar to pahoehoe flows, as the lava forms in lobes,

but in this case the quenching action of deep water retards the budding of new flows. The new lava buds then form bulbous, contracted pillow-like shapes. This process has been observed offshore Hawaii (Figure 5.9). In the rock record, pillow lava is characterised by its shape and the radial cooling joint patterns that develop as the flows contract. Commonly a very glassy rind forms around the edge of the pillow as it is quenched on contact with water. In the field, examples show the preferential weathering of these rinds (Figure 5.9). Between pillows there may be distinctive sediments with **radiolaria** (marine fossils), **cherts** or iron and manganese oxide-rich deposits known as **umbers**, which may be a product of **black smokers**.

Fragmentation of lava in water creates a deposit, **hyaloclastite**. This is a brecciated deposit with angular to flat fragments sized from a millimetre to a few centimetres containing much fresh, glassy material. Hyaloclastites can be quite complex, as they often contain both primary fragmented material and secondary, re-sedimented or reworked material within the same deposit. Where lavas flow into open water, for example into the sea, lava deltas of hyaloclastite build up (Figure 5.9) and can include mixtures of hyaloclastite, reworked **volcaniclastic sediments** and pillow sequences interbedded as prograding wedges of material that fill in the water depth as they build out. If the production of lava is high, then the delta will build out to sea. The resultant deposits can be impressive, with the largest hyaloclastite delta deposits being several hundred metres thick, as seen in Greenland and in offshore seismic images.

Cauldrons of fire: lava lakes

There are only a very few active lava lakes on the Earth. These bubbling vats of magma are very surprising; they are often long-lived and resemble a bubbling cauldron. They are characterised by an open lake surface that constantly simmers with convecting magma as it cools. This cooling process causes a viscous crust similar to that on a porridge pot that has been left on the hob too long; it ripples, breaks and subducts around the sides of the pan and between the convecting upwellings. The lakes themselves can fill and drain over time. During periods when the lakes have filled and spilt over, lava flows adorn their flanks, some flows being fed several kilometres away from the source lake. The exact geometry of the lava lakes at depth is not known. Why some are able to maintain an open magma surface is somewhat of a mystery, as there must be a fine balance of new hot magma supply coupled to constant recycling or drainback of the colder lava. When this balance is not maintained the lava lake may overflow, as described above, subside to lower levels, or ultimately switch off and harden as it cools. If the lake is breached, for instance by faults developing in the sides of a volcano, then it can drain very quickly, often with fatal results. An example of this occurred on 10th January 1977 from the lava lake at **Nyiragongo**, Democratic Republic of the Congo. Cracks appeared as the crater walls were fractured, resulting in a spectacularly quick draining of the lake. This lava lake drained in less than an hour, with outflow speeds of up to 60 mph, which devastated everything in its path. At least seventy people were killed and livestock and housing was affected. Such hazards are not common, but in the case of Nyiragongo, it is situated close to populated areas including the town of Goma.

Reportedly one of the oldest lava lakes on the planet is that of **Erta Ale** volcano in Ethiopia (Figure 5.10). The 'gateway to hell' to the local Afar tribes, it has been bubbling away for over 100 years. Erta Ale was the site of a recent expedition to capture the first full three-dimensional survey of a lava lake using laser ranging technology. **Mt Erebus**, Antarctica has had an active lava lake with

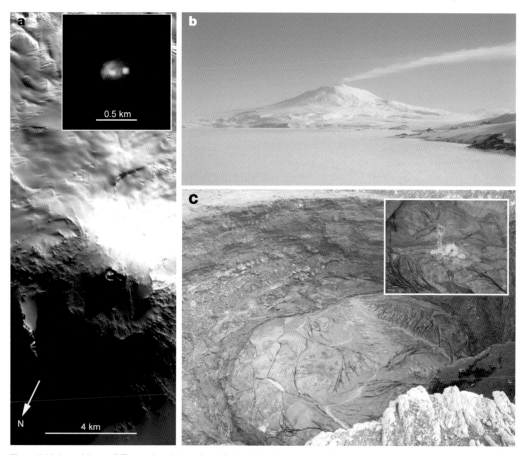

Figure 5.10 Lava lakes. **a)** Thermal emission from Erebus' lava lake, Antarctica, detected by Hyperion onboard NASA's Earth Observing-1 spacecraft (image Courtesy of NASA/JPL). **b)** Volcanic plume trail from Mt Erebus, Antarctica. **c)** Erta Ale lava lake, Ethiopia.

modern reports from at least 1972. The volcano was viewed as active from a distance in 1841 by James Ross, who named it *Erebus* after one of his ships. A degassing plume can often be seen emerging from the summit of the volcano (Figure 5.10). The lava lake can have minor explosive eruptions when big bubbles breach the surface, and it contains remarkably large **anorthoclase** crystals. This suggests the long-lived nature of the magma, enabling the growth of such large crystals. Other examples of lava lakes include Halema'uma'u, in Hawaii; Villarrica volcano in Chile; and Massaya volcano, Nicaragua.

Odd flows and weird compositions

Some of the more extraordinary types of flow occur rarely on the Earth's surface and on other planets. In the Earth's past some very hot, very runny magmas occurred, predominantly in the **Archaean**, more than 2.5 billion years ago, which are preserved today in a few old **cratons**, the oldest parts of the continental crust, found in places such as Australia, Canada and Africa. These ancient magmas are called **komatiites** and are characterised by very high magnesium contents. The eruptions that formed these flows may have been as hot as 1600 °C and would have erupted at white heat instead of the common orange glow known from eruptions today. Komatiites are commonly associated with ore deposits and have been mined for metals including platinum and nickel. A type of volcanic and intrusive rock that contains more than 50% carbonate minerals, known as **carbonatite**, is found in some places around the world. Ol Doinyo Lengai volcano, in the Great Rift Valley of Africa, is the world's only known active carbonatite volcano, which erupts small volumes of very low viscosity

lavas. The world's lowest subaerial volcano, **Dallol** in Ethiopia, rises gently from 100 m below sea level in the Danakil desert. It erupts a strange mixture of magma and salt as the volcano interacts with a thick salt basin. This dynamic interaction with subsurface salts and groundwaters results in spectacular acid pools and weird volcanogenic features (Figure 5.11).

Mud volcanoes represent a strange eruptive phenomenon that is not driven by magmatic rock, but whose features at the Earth's surface closely resemble those seen at volcanoes. Their behaviour is not very different from the low viscosity eruptions seen with carbonatites. Azerbaijan and its Caspian coastline contain some 400 mud volcanoes, more than half the world's examples. Mud volcano eruptions are driven by methane, carbon dioxide and nitrogen erupting as a slurry or flow of muddy liquids containing heated water and hydrocarbon fluids, which are frequently acidic or salty, and can be

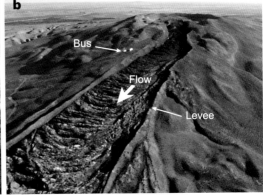

Figure 5.11 Unusual types of flow. **a)** Strange volcanic forms and sulphurous pools on Dallol Volcano, Ethiopia; **b)** View across Kotyrdag mud volcano, Azerbaijan (Sverre Planke). Note the similarities with lava flows including ropy pressure ridges indicating flow, flow channel and levees (cf. Figure 5.4).

small to truly volcano-size in scale (Figure 5.11b). Further afield, possible mud volcanoes have been identified on Mars, and strange volcanoes have been witnessed on Jupiter's moon Io which erupt sulphur, sulphur dioxide and silicate rock or lava at temperatures thought to be in excess of 1500°C (*see* Chapter 3). Some of the other outer moons of our solar system erupt water volcanoes that freeze on eruption, known as **cryovolcanism** (ice volcanism), for example on Jupiter's moon, Europa.

6 Explosive pyroclastic eruptions and their deposits

The archetypical view of a volcanic eruption is that of a sequence of explosions of molten rock ejecting material into the air and causing panic and havoc on the ground. Such explosions are commonly known as 'pyroclastic' as they involve fragments of hot juvenile clasts of magma, pyroclasts, which are ripped apart during the eruption. Unlike the effusive eruptions of lava flows, explosive volcanism can represent a more direct hazard to people and can be somewhat less predictable in terms of its timing and the volumes of erupted products. Explosive eruptions can occur in all volcanic settings, from the fire fountains that can feed basaltic lava flows to the massive Plinian eruptions that can destroy mountains, as well as lateral blasts that devastate large tracts of land (Figure 6.1). In the extreme case, the eruption of a supervolcano, involving explosive eruptions of more than 1000 km^3 of magma, would have far-reaching effects. In Chapter 4 the types and range of explosive volcanic styles helped define different types of volcanic eruption, such as Vulcanian, Strombolian, Plinian, etc. The relationships between the eruption and the amount and intensity of the events may be classified in terms of the Volcanic Explosivity Index (VEI) (Chapter 4). But what of the processes of explosive eruptions and their resultant deposits? In this chapter a closer look is taken at the components of an explosive eruption. The processes that are taking place as material is ejected from a volcano are explained, and the pyroclastic deposits that are left in the aftermath of a volcano's eruptive episode are examined.

Bombs to ash – the basic ingredients of pyroclastic rocks

To produce a pyroclastic rock from a volcanic eruption, key ingredients are required. Differing mixtures of these ingredients result in a particular type of pyroclastic rock. This view of looking at the key components of pyroclastic rock is very important, especially when we try and analyse old deposits to work out how the eruption took place and gave us what we can observe.

The major components of any pyroclastic rock may be divided into two major types of material. First, those related to the magma involved in the eruption itself, known as juvenile components or fragments. Secondly, older parts of the volcano and surrounding country rock that have been broken up and incorporated into the eruption; these are known as **lithic fragments** or **xenocrysts** where individual crystals are incorporated. Lithic fragments range from bits of other lavas and pyroclastic rocks from earlier eruptions of the volcano to a whole range of country rock types that the magma has plucked off on its way to the surface: some from shallow depths and some from very deep in the volcano's plumbing system. Diamonds are an extreme example of this, being xenocrysts that are brought up in kimberlite eruptions. Lithic fragments can be very important, as they can provide information about how violent the eruption may have been; they can be used to characterise collapse events in the walls of the old edifice, and may provide important constraints on

the plumbing system of the volcano and any contamination that may have changed the composition of the juvenile magmatic components.

In order to characterise the large variation that is seen in pyroclastic rocks, classification schemes have been developed that use the different components of the deposits and the grain size of the units. A summary of these is presented in Figure 6.2. In such schemes the term 'ash' is applied to modern

Figure 6.1 Examples of explosive eruptions. **a)** Eruption of Stromboli at night. **b)** Sunset on the 22 July 1980 eruption of Mt St Helens, looking north-east (USGS). **c)** Lava fountain feeding lava flow at the Holuhraun 2017 eruption in Iceland (Morgan Jones). **d)** A picture of Russia's Sarychev Volcano, located in the Kuril Islands, erupting, as seen from the International Space Station (NASA).

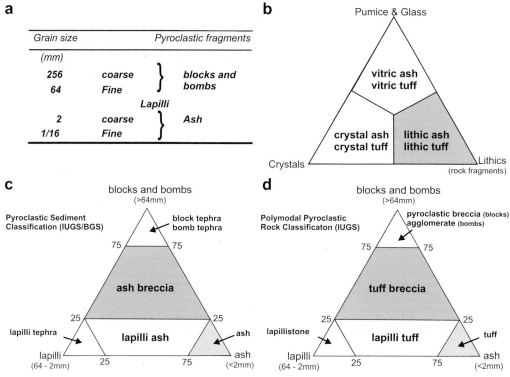

Figure 6.2 —Pyroclastic classification schemes. Simplified classification schemes based on type and size of clasts found in modern and ancient pyroclastic deposits.

unconsolidated deposits and the term 'tuff' (from the Italian 'tufo') for ancient deposits. Additionally, the term 'tephra' (of Greek origin) is used to describe pyroclastic accumulations forming unconsolidated deposits; for example, **lapilli tephra** (Figure 6.2).

The juvenile components of a pyroclastic rock warrant detailed exploration, as in many ways these are the driving forces behind the eruption and ultimately help define and classify the type of volcanic eruption forming the pyroclastic rock. Primary magma can break up into a large range of juvenile components when it is involved in an explosive eruption,

from bombs and blocks several metres in diameter to fine fragments of **volcanic ash** a few microns in size. Figure 6.3 provides a way of looking at the variety of fragments by plotting the types of eruption and the size and character of their associated volcanic clasts. For example, volcanic bombs differ from volcanic blocks in that their shape records fluidal surfaces (bombs were juvenile liquid magma when erupted). The type of juvenile component of an explosive eruption will vary with the amount of **volcanic gas**. Bombs and blocks of juvenile magma will be found where very little gas is preserved in the clast.

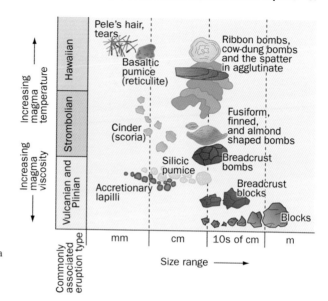

Figure 6.3 Size and types of volcanic clasts. Shapes, sizes and types of volcanic clast as a function of the type of eruption/eruptive conditions (after John P. Lockwood and Richard W. Hazlett, 2010).

Fragments filled with evidence of gas, and that are so full of vesicles that they can float, are called **pumice**. These can be found as 'frozen' magmatic froth or foam that records the vesiculation and exsolving of gas during the ascent and eruption of the magma. When vesiculation becomes extreme it breaks the magma up completely into fragments, tiny shards of magma, providing fine ash particles. Any crystals that were growing in the magma are then liberated as individual crystal clasts. The whole range of juvenile magmatic components includes particles with different densities and properties, which in turn are found or concentrated in different parts of an eruption. For example, bombs are found close to a vent, with smaller particles being found further away, and very fine ash possibly being dispersed great distances from a volcano. Some examples of volcanic products are shown in Figure 6.4.

Raining from the sky – fallout

The simplest of the primary pyroclastic processes is that of fallout, where material drops out of the sky from a volcanic plume that has been blown away from the eruption. Of course it is more complicated than that, as the ferocity of the eruption will dictate how large are the particles erupted high into the air. The winds blowing during an eruption and the turbulent winds caused by the eruption will all play a part in the distribution of falling material. In general, other than very light pumice, it is harder for clasts that are more than a few tens of millimetres in diameter to stay in the air for any great length of time. As such, fallout from an eruption column is usually restricted to pumice, fine **lapilli**, ash and small crystals (*see* Figure 6.2 for size classifications).

Material can rain out of the sky, much like snow, during the fallout from an eruption. At

Figure 6.4 Photos of different juvenile components, bombs, etc. **a)** Accretionary lapilli, Santorini, Greece. **b)** Pelee's hair, Erta Ale summit, Ethiopia. **c)** Bomb sag/impact in volcanic sediments, Santorini, Greece. **d)** Bread-crust bomb, Mt St Helens, USA. **e)** Large bomb/block, Stromboli, Italy. **f)** Bomb through broken crust of lava, flanks of Erta Ale, Ethiopia.

any stage the size range of particles falling will be very well sorted by **aeolian** processes. Size changes may occur as the strength of the eruption changes; the availability of juvenile frothy pumice and the type of suspended particle will then vary. In the resulting deposits of fallout, well-sorted grading between coarse and fine units can often be observed.

How do pyroclastic eruptions flow?

A large range of pyroclastic rocks are deposited by pyroclastic density currents (originally termed pyroclastic flows). These are turbulent mixtures of coarse particles, ash and gas. The range of different volumes and concentrations of particles, ash and gas

involved in any one flow can change during a single flow event. This results in a great range of different types of deposits. A pyroclastic density current is, in essence, a turbulent mixture of all of the pyroclastic components, from juvenile to lithic, already described, and gas. These currents flow downhill very much like fluids. They are termed low density, or low concentration, flows when there is a low ratio of gas to fragments and high density, high concentration, flows when there are far more fragments than gas. In fact there is a continuum between the two forms, and during a single pyroclastic current event it may go through stages of being low concentration and of being high concentration. These stages

Figure 6.5 Schematic summary of a pyroclastic density current. As a pyroclastic density current descends from the volcano a deposit aggrades upwards from the base, with a boundary layer defining the deposit from the moving cloud of particles. This so-called 'aggradation' can take place over minutes to hours, and even days, for some of the biggest eruptions. Photo is of a pyroclastic density current from the 7 August 1980 eruption on Mt St Helens. View from Johnston Ridge (USGS) – note photo reversed to match graphic inside of an aggrading pyroclastic density current.

will be reflected in the final deposit. Lower concentration pyroclastic density currents are often termed '**pyroclastic surges**', whilst the higher concentration versions are most often referred to as 'pyroclastic flows'.

A pyroclastic density current can originate from the fountaining or collapsing of an eruption column. Other examples can occur from the collapse of a dome; this particular type of flow is often called a '**block and ash flow**' and lacks pumice fragments; or from the lateral blast of an eruption. Importantly, a pyroclastic density current will travel across the Earth's surface and will both deposit material from its base through time and transport other material that it picks up at its base. These flows can thus accumulate through the duration of the eruption,

building up deposit as it progresses, and this can vary both laterally and vertically depending on the conditions of flow and particle concentration at any one time or place. This is schematically highlighted in Figure 6.5, and is a fundamental reason why there is such a variety of structures in pyroclastic deposits resulting from such a flow. Throughout the duration of an eruption, pyroclastic currents may wax and wane in terms of their energy and particle concentration and can bring different materials to the same locations as the eruption proceeds. When such flows are observed now, it is evident they have a sustained period of eruption and deposition. So the products of pyroclastic eruptions represent deposits that have built up over a protracted period of time lasting several minutes, hours or even days, in some of the bigger eruptions.

Understanding the deposits

Explosive eruptions will invariably produce a range of different deposit types, from those that fall from ash and light pumice clouds, which blanket the topography and are known as **fallout deposits**, to those that are formed from pyroclastic density currents. The latter can vary in their coverage of the topography, depending on the concentration of the

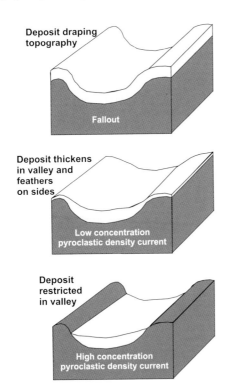

Figure 6.6 Topography of eruptive products. The different ways in which deposits cover pre-existing topography, formed from the different components of an explosive pyroclastic eruption.

pyroclastic density current (Figure 6.6). The resultant deposits of pyroclastic flows are most commonly known as **ignimbrites** and can display a range of structures in a typical deposit.

As the deposits themselves can originate from a variety of different types of lateral flow, turbulent currents and dramatically changing flow regimes, sedimentary structures are an important part of the deposits that form from explosive volcanic eruptions. Different styles of cross-bedding, grain-size sorting and grading within bedded units can occur and need to be carefully looked at throughout the deposit. These can provide a wealth of information about the style and variation of the flow regime during deposition; this, in turn, can be used to infer what may have happened at that stage in the eruption. For example, a raft of large pumice clasts that appear halfway through a deposit may indicate that the flow has waned at that stage and allowed lighter clasts to deposit or that the flow has changed direction and the point of deposition is now towards the edges of the main flow. A sudden increase in lithic fragments could mean that part of the walls of the vent have collapsed, providing this additional material, whilst the presence of cross-bedding in moderate to fine-grained deposits may indicate deposition from a low particle concentration pyroclastic density current.

Layers and structures in ignimbrites

With all of this variability (and complicated nomenclature) what hope is there of understanding the deposits of pyroclastic eruptions? Well, through careful inspection and by looking at a number of classic examples, it is possible to come up with a simplified version of what may be expected in a profile through an ignimbrite deposit. The general structures of a typical ignimbrite deposit can be summarised, and this may be used to infer what happened during the eruption that produced the deposit. In many examples, only the products of an eruption are present, and we must use these to provide valuable information about the volcanic processes that formed them. This can be of great importance in better understanding a volcano's history and its potential future hazard.

A generalised section through an ignimbrite deposit is given in Figure 6.7. This shows the types of vertical variations displayed by many deposits. Not every deposit will have all of these features preserved. Commonly, at the base of an ignimbrite deposit, there is a fallout layer of well-sorted pumice that has rained out of the Plinian cloud before the pyroclastic density currents arrived. The main deposit starts with some low density currents, which can show cross-bedding features with reverse grading of coarser material being added as the concentration of the current increases. The main body of the ignimbrite can be massive in part, but it will show some grading features as the currents waxed and waned. It may have lithic or pumice layers in the deposit, as described previously, and it may show a general increase in pumice clasts towards the top of the deposit. The upper parts of a deposit are the result of a waning of flow and current concentration, leading to finer cross-bedded structures, which are commonly overlain by **ash-fall deposits**. It must be stressed that this is an idealised example, and in any one eruptive episode there are

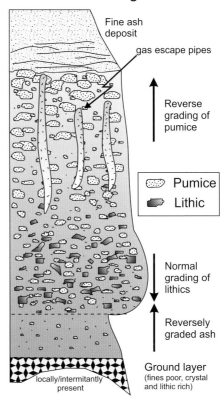

Figure 6.7 Schematic through an ignimbrite deposit. A typical ignimbrite is made up of layers depicting the different stages of the eruption and the build-up of the deposit (as depicted in Figure 6.5). The lower layers occur due to the first blast of the explosive eruption with the main part of the deposit formed by the pyroclastic flow as it flows along and the deposit aggrades upwards during deposition. Commonly you can find a concentration of coarse pumice clasts towards the top, and fluid escape pipes as trapped gases have been escaping from the deposit. The upper parts are characterised by fine ash that settles as the flow ceases (After Sparks *et al.*, 1973). (Note these structures are idealised and may not occur in all ignimbrite sections, and can be repeated through deposits formed from single eruptions where the flow has changed direction and strength during the build-up of the deposit).

Figure 6.8 Photo examples of ignimbrites. **a)** Thick ignimbrite sequence (white) from the Minoan Eruption, Santorini, Greece. **b)** Ignimbrite deposit at the coast, Tenerife, Spain.

likely to be modifications to this, as deposits from eruptions having several pulsed events will show more internal complexity than those from single events.

Welded ignimbrites

The process of erupting juvenile hot magma in an explosive eruption is a complex interplay of new magma, fragments of existing rock and debris, and can involve eruption mechanisms that lead to ignimbrite deposits showing a variety of structures. A further complication to the types of structures that can be preserved in an ignimbrite are the processes that occur when the erupted material is particularly hot as it comes to rest on the ground. This can produce a group of rocks known as **welded ignimbrites** (Figure 6.8). In such circumstances the juvenile fragments remain hot enough during deposition to start to fuse together. They then begin to revert to

Figure 6.9 Example of a welded ignimbrite. Photo highlights flattened, flame-like pumice clasts ('fiamme') giving a streaked-out eutaxitic texture and blocky undeformed lithic clasts, welded ignimbrite, Bushop Tuff, California, USA.

their original magmatic form. In very extreme circumstances the whole deposit can start to re-form into structures more commonly associated with lava flows, known as **rheomorphic ignimbrite**, but these are rare. More commonly in the middle to lower parts of the ignimbrite, where the material retains heat the longest, the weight of the overlying material helps to weld and deform the original depositional structures. The light, frothy pumices that are caught up in the deposit start to flatten and streak out and form flame-like structures which are known as **fiamme** (after the Italian for flame). The background, ashy, material sticks together and hardens the ignimbrite, and a texture develops where ductile juvenile objects are flattened and streaked out parallel to the long axis of the deposit, producing what is called a **eutaxitic texture** (Figure 6.9). In intensely welded examples, parts of the deposit become so much like the original magma that they can flow, creating folding structures.

The process of explosive eruption and transport of pyroclastic material will normally significantly cool and harden the juvenile magmatic component, such that, where welded or partly welded ignimbrites are found, certain conditions will have occurred to help promote their welded character. For example, it may be that the eruption was particularly large so that all the heat could not be dissipated rapidly. Welding can occur in parts of deposits that are ponded in pre-existing topography or are close to the source of the eruption. This can also happen where the component of juvenile material makes up almost all of the eruption, and where the magma coming out of the volcano is superheated.

Explosive hydrovolcanic eruptions

Hydrovolcanic or phreatomagmatic eruptions occur when magma interacts with water or ice. They can occur when lava flows enter the sea or lakes and also when volcanoes erupt under water, when emergent volcanoes break the surface of the water, and where volcanoes erupt under ice. The interaction of a volcano with groundwater can enhance the explosivity of the eruption. There are a whole range of interactions that depend on the relative ratios of water to magma, the temperature and viscosity of the magma, and how charged with gas the magma is initially. One of the main results of adding water to eruptions is that they become more explosive and violent. Vulcanian and **Phreatoplinian** eruptions occur when significant additional water is involved in very explosive eruptions. Where shallow magma comes into contact with groundwater to cause relatively small single explosions, circular eruption craters with shallow deposits known as **tuff rings** occur. A specific type of volcanic fragment known as **accretionary lapilli** can be found in deposits where eruption columns contain moisture (Figure 6.3). These are balls of sequential coats of ash sticking to small objects and growing in concentric rings, a little like volcanic hailstones.

Lava erupting under water can produce pillow lavas, but in other examples it fragments violently. Submergent volcanoes provide one of the most spectacular examples of these eruptions; as they breach the water surface they start to erupt walls of water and lava fragments. These provide an example of how new land can be formed on the planet. A good example is the eruption of

Figure 6.10 Eruptions through and beneath the sea. **a)** Surtsey seen from the air on 30 November 1963, 16 days after the beginning of the eruption. (U.S. National Oceanic and Atmospheric Administration); b–d) Underwater sampling of active volcanoes taken from ROV (remotely operated underwater vehicle): Jason II, Lau Basin, south-west Pacific, 2009 (Marine Geoscience Data System).

Surtsey in Iceland (Figure 6.10). Surtsey was formed in a volcanic eruption that began 130 metres below sea level and reached the surface on 14th November 1963, when it broke through the water with spectacular eruptions. This continued with the formation of a new island until 5th June 1967, when the island reached its maximum size of 2.7 km². It is now a special reserve dedicated to researching how plants and animals colonise new land, and a UNESCO World Heritage Site. Other examples include the pop-up island of Ferdinandea off southern Sicily, which periodically emerges from the Mediterranean.

Reworking the explosive products

When volcanoes leave deposits that are fragmented loose material, such as unconsolidated pyroclastic material, this can soon be remobilised and resedimented to form new deposits, which differ in whole or in part from their original volcanic origins. One of the most commonly reported resedimentation phenomena is that of lahars, flows rich in particles and water, often called mudflows, which are very hazardous, as they can move very quickly and destroy almost anything in their path. Lahars can occur during eruptions, particularly where ice on a volcano is rapidly melted to provide an

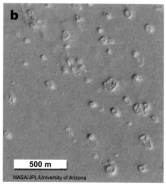

500 m

NASA/JPL/University of Arizona

Figure 6.11 Examples of rootless cones. **a)** Classic view of rootless cones at the edge of Lake Mývatn, Iceland (© Inbound Horizons/Shutterstock). **b)** Rootless Cone Field near Aeolis Planum, Mars (NASA/JPL/University of Arizona).

abundant source of water. Hazards due to lahars can occur months or years after a new eruption, as unconsolidated volcanic debris can easily be remobilised by heavy rains; for volcanoes in the tropics this is a particular hazard. Reworked volcaniclastic rocks, where less than 25% of the material is of primary volcaniclastic origin, are termed **epiclastic deposits**, and the evidence of rounding of clasts by transport, and subsequent secondary sedimentary structures, can be used to help distinguish these from primary volcanic products.

Explosions with no roots

An interesting and somewhat enigmatic type of explosive eruption can occur where there is no magmatic feeder directly beneath the vent. This type of eruption is caused when lava flows over water-saturated ground, snow or ice. These are termed 'rootless cones' after the pitted eruption cone that is left after such an event. The hot lava rapidly heats the underlying water and turns it to steam, which involves rapid expansion and a dramatic increase in pressure. If the pressure builds up enough the steam can burst through the partially molten lava above, causing an explosive eruption, and a resultant tuff and spatter ring around the vent. Classic examples of these are found in Iceland (Figure 6.11a), and such features have also been interpreted on the surface of Mars (Figure 6.11b). On 16 March 2017, a BBC crew filming on Mount Etna were caught in a rootless eruption from a lava flowing over snow.

7 Igneous intrusions – a window into volcano plumbing

When a modern volcano, erupting hot lava at the surface, is observed the complex plumbing beneath the volcano is masked. Much like the complex plumbing behind the taps in our houses, the routes by which the magma travels from the magma chamber below to feed the volcano at the surface are a vital part of the volcanic system – its plumbing. There are a large variety of ways in which magma gets from its origins to the surface. When it gets trapped below by cooling in the crust, or if it never makes it to the surface, **igneous intrusions** occur. In order to fully categorise the types of intrusions in volcanic systems, the system as a whole needs to be looked at (Figure 7.1). There are a number of examples on the Earth where key elements of the plumbing system are preserved and exposed by erosion. In and around recent volcanoes, where erosion has revealed the shallow system, and with deeply eroded old volcanoes, the key to the puzzle is to unravel the three-dimensional shape of the intrusion and how it relates to the pre-existing country rocks it cuts through. Deeper into a volcanic system, outcrops are needed that expose the middle and lower parts of the crust and, in some cases, the mantle parts of the lithosphere. Thus evidence may be found of intrusions that are part of the deep plumbing of the volcanic system, from the very source of the melt itself.

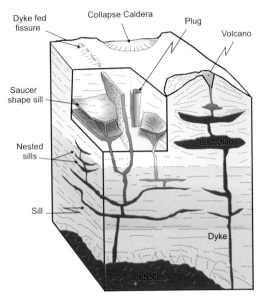

Figure 7.1 Schematic of a volcano plumbing system. Volcanoes are fed by a complex network of underground intrusions that make up the magma plumbing system.

Sills and dykes

The simplest forms of intrusion that are associated with the plumbing of volcanic systems are tabular sheets known as sills and dykes (*American usage = dike*). In general, a sill is a tabular sheet of igneous rock that is concordant with the bedding or major compositional layering of the rocks that it cuts through. Sills are often horizontal to sub-horizontal. A dyke, however, cuts the bedding of the host

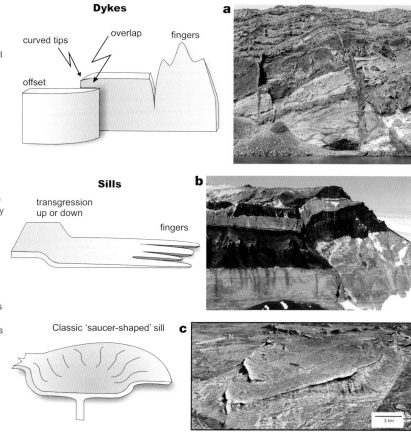

Figure 7.2 Simple sketches of dykes and sills with examples. **a)** Dykes: vertical to sub-vertical sheets of magma that cross-cut the country rocks. The example given shows a dyke network exposed on the walls of the caldera at Santorini volcano, Greece. **b)** Sills: horizontal to sub-horizontal sheets of magma that are often concordant with bedding within country rocks. The example given shows thick (c.200 m) dolerite sills (dark brown) within sedimentary rocks (yellow), in the Dry Valleys, Antarctica. **c)** Saucer-shaped sills found in sedimentary basins and sometimes within volcanic stratigraphy. The example given is the Golden Valley Sill, Karoo, South Africa (after Nick Schofield, 2008).

rocks at an angle, so it is discordant, and the sheet is often vertical. This simplified ideal is shown in Figure 7.2, but it should be noted that all forms and varieties of sheet intrusions can be found, so in some instances it may be difficult to decide whether to call something a sill or a dyke, or they may be transitional between the two. An example where there are clear transitions from one form to another is found in intrusions known as saucer-shaped sills (Figure 7.2c). These

bodies have planar sill-like parts which then ramp up around the edges, cutting through the country rock like the rims of a saucer. Examples of these types of intrusion are found in sedimentary basins such as the Karoo in South Africa, and can be imaged in three-dimensional seismic surveys of offshore sedimentary basins such as the North Atlantic Margin (Figure 7.3). This relatively new technique of three-dimensional imaging of igneous intrusions in offshore settings is

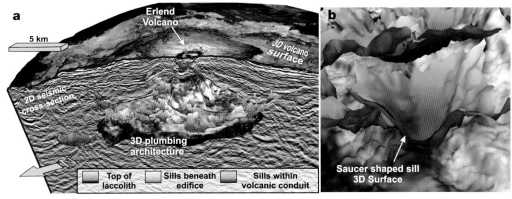

a

5 km

Erlend Volcano

3D volcano surface

2D seismic cross-section

3D plumbing architecture

| Top of laccolith | Sills beneath edifice | Sills within volcanic conduit |

b

Saucer shaped sill 3D Surface

Figure 7.3 View into the 3D plumbing system beneath a ~57 million year old volcano using 3D seismic imaging. **a)** The main feeder system is multi-tiered with a main laccolith complex and associated sills both beneath the volcanic edifice and within the volcanic conduit. **b)** Close-up image of one of the sills reveals a classic saucer-shaped geometry (see also Figure 7.2), (images by Faye Walker with data courtesy of TGS; see also 'Inside the volcano', by Walker et al., *Geology*, 2020).

providing new insights into the complexity of some igneous plumbing systems.

With detailed inspection of intrusions such as sill systems in outcrops, it is possible to decipher emplacement processes and directions. This is often enabled by the presence of lobe or 'finger'-like structures (Figure 7.4), which can indicate flow directions and linkages in the third dimension. These types of observations are made on a number of scales from 3D air surveys with drones to detailed outcrop mapping of structures.

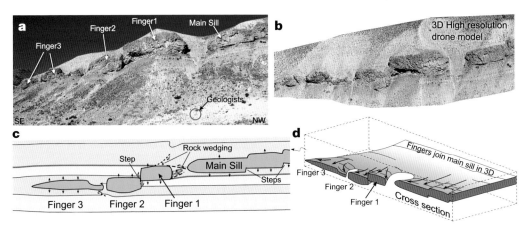

a Finger1 Finger2 Finger3 Main Sill Geologists SE NW

b 3D High resolution drone model

c Step Rock wedging Main Sill Steps Finger 3 Finger 2 Finger 1

d Fingers join main sill in 3D Finger 3 Finger 2 Finger 1 Cross section

Figure 7.4 Detailed look at magma fingers in outcrop at Las Loicas, Neuquén basin, Argentina (Olivier Galland). **a)** Cliff view of fingers. **b)** 3D outcrop model generated by drone. **c)** Detailed interpretation of the fingers and emplacement indicators. **d)** Idealised 3D model of the situation at Las Loicas highlighting relation to main sill.

Sills and dykes are often intricately associated with the volcanic systems that they feed and, in many examples, can be found intruding into and cutting through erupted volcanics in the same system. Some of the most dramatic sills can be up to 300 m thick, for instance the examples from the Dry Valleys in Antarctica (Figure 7.2b). Where dykes occur in great numbers they are termed **dyke swarms**, where they commonly occur as aligned stripes cutting through the country rock (Figure 7.5). They can occur in a number of settings, but the phenomenon is most commonly preserved in parts of the planet where rifted margins are associated with massive outpourings of lavas known as flood basalts. Some of the most spectacular images of dyke swarms have been achieved through magnetic surveys. Depending on the nature of the surrounding country rocks, these may be able to pick up the relatively iron-rich nature of the dyke rocks. Figure 7.6 highlights the different cross-cutting dyke swarms that have been imaged across Northern Ireland. There are at least three different populations of dykes that

Figure 7.5 Swarm of dykes as seen from a drone around Puntilla de Huincán, Neuquén basin, Argentina (Olivier Galland).

relate to the development of the North Atlantic Igneous Province, which occurred as a result of Europe splitting apart from North America some 60–55 Ma ago. Other large dyke swarms are associated with the African Karoo Igneous Province of around 180 Ma and the Mackenzie dyke swarm in Canada, 1267 Ma.

Laccoliths, lopoliths and plugs

The minor intrusions that are associated with volcanic systems include bodies known as **laccoliths**, **lopoliths** and **plugs**. Essentially, these are relatively small bodies of hardened magma, but all have specific shapes that differ from sills and dykes. A laccolith is an asymmetrical intrusion where the roof parts of the body have been preferentially inflated in the middle and less on the outside, giving it a mushroom type appearance. A lopolith, on the other hand, is more or less the reverse of a laccolith. Here the middle part sags downward so that the intrusion is more like a bowl or upside-down mushroom. A plug

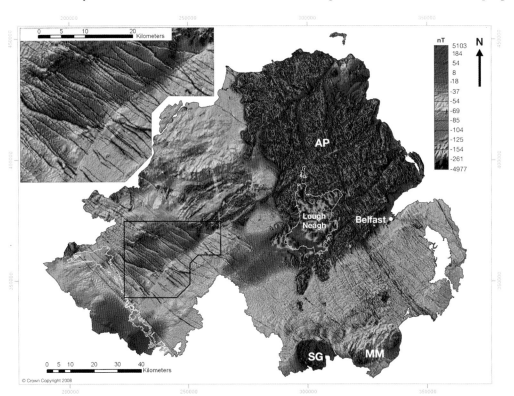

Figure 7.6 Dyke swarms in Northern Ireland. Total magnetic intensity (TMI) anomaly map (sun shaded 35/075) of Northern Ireland showing dyke swarms and prominent magnetic anomalies associated with the Antrim Plateau (AP), Slieve Gullion (SG) and the Mourne Mountains (MM). Inset map shows the offsets of dykes across the Tempo–Sixmilecross Fault and the Omagh–Tow Valley faults (Colour scale bar is in nanotesla (nT)) (Cooper *et al.*, 2012).

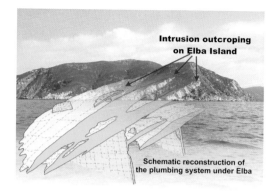

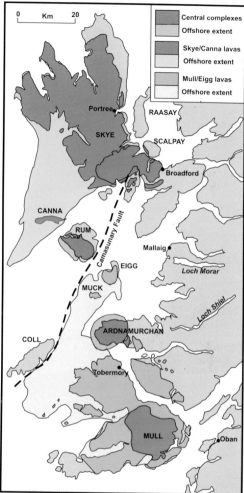

Figure 7.7 Volcano plumbing on the Isle of Elba, Italy. Interpretation of the intrusion network underneath the Island of Elba, Italy, extrapolated from the onshore exposures, forming a complex plumbing system (Sergio Rocchi).

represents a cylinder-like body of hardened magma. Schematic examples of these are given in Figure 7.1. Studies of some complexes that are made up of a number of small intrusive bodies incorporate recognised nested laccolith and lopolith sequences such as those in the Henry Mountains of Utah, USA, and on the island of Elba (Figure 7.7). Here the intrusions can occur one above another as though they were being fed through the same system in the shallow parts of a volcano's plumbing. Laccoliths, lopoliths and plugs are generally circular in plan view, and can have multiple stages of intrusion.

Fossil magma chambers

The thought of a magma chamber, a hot volume of molten rock sitting beneath a volcano, is not so far-fetched. Often magma ponds in the shallow crust in structures called magma chambers. These can feed volcanoes above them through sills, dykes, plugs and vents (Figure 7.1). But what does a magma chamber really look like, and how do

Figure 7.8 The Scottish volcanic centres. Simplified geological map of the location of the Scottish volcanic centres (central complexes) of Skye, Rum, Ardnamurchan and Mull, with their associated lavas (Adapted from Emeleus & Bell, 2005).

they form? There are a number of fossil magma chambers that are now exposed at the surface through erosion. These allow us to explore the insides of these old volcanoes,

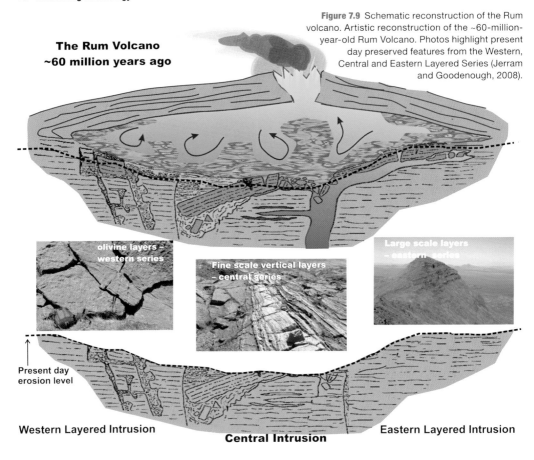

Figure 7.9 Schematic reconstruction of the Rum volcano. Artistic reconstruction of the ~60-million-year-old Rum Volcano. Photos highlight present day preserved features from the Western, Central and Eastern Layered Series (Jerram and Goodenough, 2008).

The Rum Volcano
~60 million years ago

olivine layers – western series

Fine scale vertical layers – central series

Large scale layers – eastern series

Present day erosion level

Western Layered Intrusion

Central Intrusion

Eastern Layered Intrusion

something clearly impossible with modern 'live' volcanoes. Some of the best examples of fossilised volcanoes are found in the British Palaeogene (British Tertiary) volcanic province in NW Scotland, on the islands of Skye, Rum, Mull and on the Scottish mainland in Ardnamurchan (Figure 7.8). The volcanic centres of Skye and Rum are particularly well exposed and played a vital part in the recognition of the link between volcanic and intrusive rocks. Early work at the start of the 1900s started to map out the cross-cutting features preserved in these centres. It soon became clear that they represented the remnants of ancient magma chambers that had formed magma reservoirs in the shallow crust, feeding Tertiary volcanoes. The chambers themselves were originally thought of as big balloon-like bodies of magma, as their surface expression was somewhat circular, and many sketches and diagrams in textbooks show these balloon-like geometries.

However, more detailed mapping and analysis has revealed a very complex multiple intrusion network that makes up these igneous centres. The chambers themselves were never as thick as their final deposits – often kilometres thick – and, instead of a big balloon, they formed large laccolith-like bodies, a hundred metres or so thick but laterally extensive. These thin chambers acted as temporary reservoirs for successive magma batches as they were on their way to the surface. In part the magma cooled whilst in the reservoir; this cooling process resulted in crystals growing out of the magma to form deposits on the floor of the magma chamber. Through repetitive cycles of the magma being recharged and crystals accumulating, a much thicker overall accumulation of igneous rocks resulted. This process is schematically indicated in Figure 7.9, as a complex magmatic chamber beneath the Rum volcano some 60 Ma. The igneous centres in the British Palaeogene, as well as other examples around the world, are vital to understanding the complexities that occurred beneath volcanoes in the past and which, by analogy, are occurring in shallow and deep magma reservoirs beneath contemporary systems.

The roots of mountains – massive batholiths

In some instances the amount of igneous plutonic rock that is exposed at the surface far exceeds what might be expected for simple plumbing of a volcanic system, even when some of the larger magma chambers are considered. In these instances huge tracts of land, greater than 100 km^2 but often several thousand square kilometres, composed almost entirely of intermediate to acid rock compositions such as granites, are found. They have a relatively minor mafic component. Such areas on the planet are known as 'batholiths' (from the Greek for bathos – depth, and lithos – rock). Examples of such batholiths include the Sierra Nevada Batholith in California, USA, and the Coastal Batholith in Peru. Such batholiths are popularly thought of as the roots of mountain chains, as many of the world's eroded remnants of mountains expose such large outcrops of predominantly granite.

On close inspection, a **batholith** itself is made up of many smaller intrusions or **plutons**, which collectively build up to make the mass of igneous rock. Often they are also riddled with cross-cutting sills and dykes and have many complex contact relationships between themselves and with the country rocks they intrude into. Initially these plutons were thought to rise up buoyantly as big **diapirs** (balloon-like bodies), which helped explain their often circular shape in plan section. The lighter granitic melts would cease ascent as they reached a level of neutral buoyancy, where they would collect. More recent studies, like those into the magma chambers, have started to unravel the detail behind the development of plutons that feed batholiths. A picture of dyke-like feeders that balloon out at the point of resting in the crust is more realistic, as it is physically difficult to move such volumes of material as single big diapirs. On close inspection many of the plutons themselves comprise multiple sheet-like intrusions similar to the magma chamber systems already described. The direct relationship of batholiths to the volcanic systems higher up in the lithosphere, prior to erosion, is difficult to establish, but such huge volumes of magma must have played some role in feeding shallower intrusions and volcanoes at the surface.

Plumbing the oceans

With about 70% of the Earth's surface made up of ocean crust, most of the Earth's surface is composed of volcanic rocks. The Earth's ocean crust is generated by volcanism at **mid-ocean ridges** and destroyed at subduction zones. Thus the Earth's ocean floor is no older than about 200 million years. But what feeds this massive conveyor system of ocean crust creation? Clearly it is difficult, as with any modern volcanic system, to see inside. This is even more difficult in the case of the volcanoes that make up the mid-ocean ridge systems as, for the most part, they are under significant depths of water. As with other volcanic systems, ancient examples are required,

where erosion has exposed the interior of the volcano. In a number of isolated examples on Earth, parts of the ocean lithosphere have been preserved by processes known as 'obduction', being placed on top of continental crust, so that the inside workings of the ocean crust plumbing system may be examined. These examples are known as 'ophiolites' and they are relatively rare. For the most part, ocean lithosphere is more dense than continental lithosphere, and thus it more commonly sinks below, instead of being obducted on top, of the continental lithosphere. In order for the ocean crust to be uplifted onto the continents, complex tectonic systems are required, and in many cases the preserved parts of the ocean crust, known as **ophiolite complexes**, are highly faulted and difficult to understand. In addition, they can be deformed and altered by metamorphic processes so that only remnants of

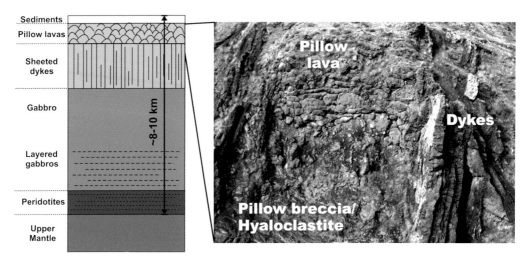

Figure 7.10 Section through an ophiolite with examples. A section through the ocean crust as preserved in an ophiolite goes from intrusions and dyke swarms at depth through to pillow lavas, hyaloclastites and sediments at the surface. The example shown is from the Troodos Ophiolite, Cyprus.

the original structures can be seen. Examples from the few well-preserved ophiolite complexes indicate a plumbing system that feeds magma from the mantle to the surface, forming a specific structure and set of rock types from deep to shallow that are characteristic of the ocean crust.

The entire thickness of ocean crust is around 8–10 km and there is not a great distance between the newly formed lithosphere and the mantle that feeds it. In mid-ocean ridge systems the hot mantle is at or near this 10 km depth. As the crust and mantle move away from the ridge, the upper parts of the mantle cool and effectively thicken the lithosphere away from the ridge. Melt collects in a sort of triangular area beneath the ridge and is focused towards the base of the crust (*see* Figure 3.2). Here the melt produces shallow magma chambers where crystallisation can occur. Layers of crystals collect at their bases, similar to those in the fossil magma chambers described earlier. These shallow magma chambers in turn feed dyke systems that further channel the magma to the surface. Lava is then erupted at the ridge itself as a mixture of pillow basalts and hyaloclastite deposits. Often the upper lava/hyaloclastite sequences are riddled with dykes and sills, which further cut them as the magma conveyor belt builds new crust and this thickens and moves away from the ridge. The standard profile of the oceanic crust representing this very efficient plumbing system is thus developed (Figure 7.10), with layered ultramafic rocks at the base, followed by sheets of dykes and, finally, with erupted material at the top. In many of the examples from the rock record, a small sliver of the lithospheric mantle is also captured when the ophiolite is obducted, and so the full transition from mantle to surface can be examined.

Ophiolites are of particular importance in terms of the minerals that are often present in their sequences. Due to the fact that the oceanic lithosphere is generated under water, it is quite different to a normal subaerial volcano. There is a constant source of groundwater available, and this is very salty. The seawater percolates down deep into the crust

Figure 7.11 a) Example of a black smoker in the Lau basin, southeast Pacific (Schmidt Ocean Institute, ROV ROPOS). **b)** Cross section through hydrothermal chimney vent chimney from East Diamante Caldera in the Mariana volcanic arc, west Pacific Ocean, rich in zinc and iron minerals (USGS).

and interacts with the volcanic plumbing system to set up huge **hydrothermal** circulation cells. These are best manifested on the sea bed by 'black smokers' and '**white smokers**' where the hot fluids are vented out through pipes in and around the oceanic ridges (Figure 7.11). These are invariably charged with a number of dissolved elements and minerals that precipitate on contact with the cold ocean bottom waters. The vents can be as hot as 450 °C, with black smokers being formed at the hotter vents and white smokers where temperatures are lower. Black smokers commonly have metal sulphides in them, which are quickly converted to sulphide deposits. White smokers contain more calcium, barium and silicon. In the ancient ophiolite, the concentrations of metal sulphides can reach economic levels as massive sulphide deposits, which have been mined for metals such as zinc, copper and lead.

Some superb examples of ophiolite rocks can be found at the Semail Ophiolite in Oman and the United Arab Emirates; the Bay of Islands Ophiolite in Gros Morne National Park, Newfoundland; and the Troodos Ophiolite in the Troodos Mountains of Cyprus (Figure 7.10). Others exist around the world, in various states of preservation, as relatively isolated examples in bigger tectonic belts. Yet it is interesting to reflect on the limited number of 'fossil' examples of ocean crust plumbing systems, as preserved in ophiolites, compared to the vast amount of ocean lithosphere there is on the planet.

Diamonds are a volcano's best friend

One specific type of volcano is worth mentioning separately in this magma plumbing chapter, as they are mostly identified by their pipe-like plumbing system, and are very special in terms of what they bring to the surface. These are the **kimberlites** and **lamproites**, which are igneous rocks that tap great depths in the Earth and bring high-pressure minerals to the surface including, most famously, diamonds. The term 'kimberlite' comes from the town Kimberley in South Africa, where a discovery in 1871 led to a diamond rush. The town today has a great big hole (Figure 7.12) left as testament to the mining of the kimberlite rock downwards over the years. Many of the early kimberlites that were discovered were found in pipe-like structures that extended vertically downwards in a carrot shape (Figure 7.12). This classic image of kimberlites has been modified over the years, as they have been found in intrusions of a variety of shapes and as surface expressions of volcanic deposits, particularly in examples from Canada. The rocks themselves are ultrapotassic and ultramafic type magmas that contain a large number of different crystals that have been incorporated along the way (**xenocrysts**), and rock fragments (**xenoliths**). The fact that they originate from significant depths in the Earth is indicated by the presence of minerals such as garnets and the diamonds for which they are famous. The volcanoes they feed at the surface map out as funnel-like constructs, a type of **maar**, which are easily eroded in kimberlite deposits and rarely preserved in the rock record, so the pipes that feed them are more commonly found. No kimberlite eruption has ever been witnessed by Man, but the great distance travelled from depth and the relatively small volume of magma involved suggest that they would be relatively short-lived but spectacular eruptions.

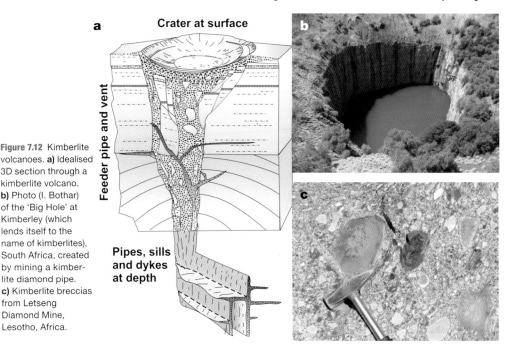

a   Crater at surface

Feeder pipe and vent

Pipes, sills
and dykes
at depth

b

c

Figure 7.12 Kimberlite volcanoes. **a)** Idealised 3D section through a kimberlite volcano. **b)** Photo (I. Bothar) of the 'Big Hole' at Kimberley (which lends itself to the name of kimberlites), South Africa, created by mining a kimberlite diamond pipe. **c)** Kimberlite breccias from Letseng Diamond Mine, Lesotho, Africa.

Imaging modern volcanic plumbing

Monitoring modern volcanoes includes a look into the subsurface in and around a modern system to find key signs of when magma nears the surface to help predict when a volcano might erupt. As with any image looking into the Earth's crust, such as seismic surveying, the amount of detail dramatically diminishes with depth. As magma moves through the lithosphere on its way to the surface, it causes tiny earthquakes. With seismic monitoring over time to record and collate these mini-quakes, a picture may be built up of the plumbing underneath some of the most closely observed volcanoes. Where this time record is extensive, a three-dimensional picture may be built up of the

structures beneath the volcano. This can be used with other data, such as from gravity and electromagnetic measurements, to enhance our understanding. Although the picture is not perfect in each case – our view is clouded by poor resolution and limited amount of data available – a virtual window into the inside of the magma system can be constructed. The volcanic plumbing under parts of Hawaii has been revealed using the available data (Figure 7.13). It is thought that partial melts collect and then start to move upwards from deep in the mantle, driven by the Hawaiian plume. They are first detected around 32 km deep, where micro-seismic activity can begin to be imaged. The magma collects in shallow reservoirs around

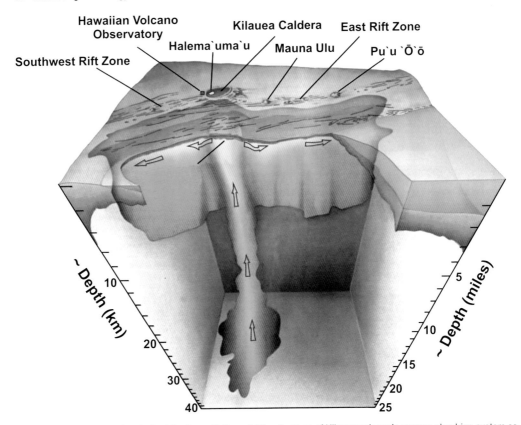

Figure 7.13 Three-dimensional plumbing beneath Hawaii. The structure of Kilauea volcano's magma plumbing system as realised using over 25 000 located earthquakes between 1969 and 1985 (After Ryan, 1988).

1.5–6 km beneath the volcano's summit, and then moves through into the shallow plumbing of the volcano. These images will be refined as monitoring of the magma movement and location is enhanced by using ever-improving technology. Ultimately there may be virtual plumbing maps for volcanoes around the world.

Strange intrusions and associations

Not all 'intrusions' are what you might expect. In fact magma is not the only thing that can intrude into a host rock; it's all about pressure (or over-pressure) fluidisation, and in many cases heat. These 'strange' intrusions can take the form of sediments that have been forced into a host rock due to extreme fluid pressures that build up during burial. These are often termed 'clastic dykes' and are thought to form rapidly by fluidised injection (mobilisation of the pressurised pore fluids), and should be distinguished from other 'dykes', which form passively by water, wind, and gravity (sediment falling into open

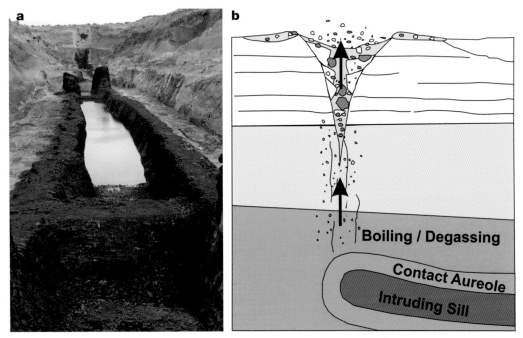

Figure 7.14 Examples of alternative intrusions. **a)** Bitumen Dyke in the Toribia mine, Neuquén, Argentina (Olivier Galland). **b)** Schematic outline of hydrothermal vent structure emanating from the tip of a sill due to heat and pressure (Sverre Planke).

cracks and crevasses within the host rock). One particularly 'strange' occurrence is that of bitumen dykes, associated with organic-rich sediments and petroleum systems. Classic examples can be found in Utah (USA) and in the Neuquén basin (Argentina). The ones in Argentina are particularly spectacular (Figure 7.14a), where bitumen-rich breccias are forced up as dykes through the host rock associated with the heating of organic-rich shales intruded with magmatic sills. In fact the intrusion of sills into different types of sediment hosts causes localised metamorphic reactions, and the resulting heat and pressure can result in the mobilisation of material in the sill's contact zones in the form of **hydrothermal** vents that punch up through the surrounding rocks, and often make it through and erupt at the surface (Figure 7.14b).

8 Volcanoes, life and climate

Volcanoes play an important role in the Earth's climate as a direct way for the Earth to de-gas. This has been the case since the formation of the Earth and is still important today. Not only do volcanoes contribute directly to the gas budget of the planet, but they also have the potential to contribute particulate matter and gases to the upper atmosphere, if the eruption is sufficiently powerful. Volcanoes have played a key role in the evolution of the Earth and even the development of life itself. Today our concern is about the potential effects of some of the largest eruptions, those of supervolcanoes, in terms of what the climatic consequences might be. In order to gain an understanding of this, the effects of modern eruptions can be used as proxies for what could happen to the climate, as the observed effects are scaled up to predict what might occur if the eruption were much bigger. Evidence from the past, when some of the largest eruptions in Earth's history took place, can be examined to see what effects were recorded in the rock record relating to the time corresponding to these big eruptions. Super-eruptions have never been witnessed, so this form of reconstructive study is very important to see how the Earth reacted to a really large volcanic event in its past. The problems lie with the shortcomings of the rock record; the older the eruption the less evidence may be preserved. This chapter considers how volcanoes may have affected and moderated the Earth in terms of climate and life. General points as to the potential climatic effects of future eruptions on our planet can be made, although there is work still to be done on this topic.

In the beginning – volcanoes and the origin of life

Volcanoes are thought to have played a very important role in the origins of life on Earth. An experiment conducted in the middle of last century aimed to mimic chemical reactions that occur in vapour-rich volcanic eruptions. The experiment circulated methane, water vapour, ammonia and hydrogen in a sealed container through which an electric spark was passed to help the mixture react. The result was the production of organic compounds. Recent reconstructions of the experiment with modern apparatus have identified 22 different **amino acids**, the building blocks of proteins. The **primordial soup**, the mix of volcanic gases on the Earth, reacting with lightning, could have initiated the development of organic compounds and thus the origins of life. Volcanoes may also have performed an important role in the development of an oxygen-rich atmosphere at around 2.5 billion years in Earth's history. This change from a **reducing** to an **oxidising** atmosphere promoted the development of more complex life on the planet. It is thought that a change to a more stable tectonic regime, involving a shift from predominantly submarine to a mix of subaerial and submarine volcanism, occurred around this time. This would have been a situation and environment more similar to that observed today. Submarine volcanoes are more reducing than subaerial volcanoes, and this rise of the volcanoes out of the sea would have reduced the overall sink for oxygen and led to an increase in atmospheric oxygen.

These are just two hypotheses about the early development of Earth that suggest that volcanoes have played a profound role in the origins of life. As volcanoes are the main outlet for gas to escape from within the Earth, it is not surprising that they must have played a role in the early atmospheric fixing of our planet. If volcanoes pump out gases, then, if there are large volume eruptions, will these have affected the Earth?

LIPs and mass extinctions

The largest volcanic eruptions have occurred irregularly throughout Earth's history. These events are known as Large Igneous Provinces (LIPs) (Figure 8.1). Individual eruptive events were of the order of 1000–5000 km^3. The majority of the eruptive products were mostly basaltic in composition, and these are often manifested as large, extensive lava flows or 'flood basalts'. When the major flood basalt events in the geological record are plotted against the known **mass extinction events**, a remarkable correlation is revealed (Figure 8.2). Key events such as the **end-Permian extinction** (~250 Ma) are coincident with one of these large outpourings, in this case the **Siberian Traps**. Major **anoxic events** such as those associated with the North Atlantic Igneous Province (stretching across the North Atlantic from Greenland to Norway) are potentially implicated in climatic

fluctuations caused by volcanism. The most widely debated of the potentially volcanically driven extinction events is that of the dinosaurs at the end-Cretaceous (~65 Ma). The **Deccan Traps** are the major contemporary igneous event that is implicated. A large asteroid impacted the Earth at around the same time, and this has led many to suggest that the dinosaur extinction was not caused wholly or in part by the climate change resulting from the activity of the Deccan Traps. However, it can be argued, convincingly, that none of the other major correlations of mass extinction with large igneous provinces have any associated asteroid impact. The relationship between large volcanic events and major changes in species looks very much like a key component of our planet's development.

It is suggested that massive volcanism caused or contributed to the end-Cretaceous, end-Permian and end-Triassic extinctions. The vast volcanic activity will have resulted in huge volumes of gases, such as carbon dioxide and sulphur dioxide, being released. These will have had dramatic regional effects on vegetation. The LIPs may even have influenced the Earth's **albedo** (the sun's light that is reflected from Earth), because of the massive area of the Earth being covered with the lava flows, and also because of the volcanic ash particles in the atmosphere. The key to developing an understanding of this relationship further is the need to better constrain the absolute ages of the volcanic events, their overall duration, and the best estimates of the volume and composition of the gases that they may have added to the atmosphere and oceans. Modern examples of eruptions, such

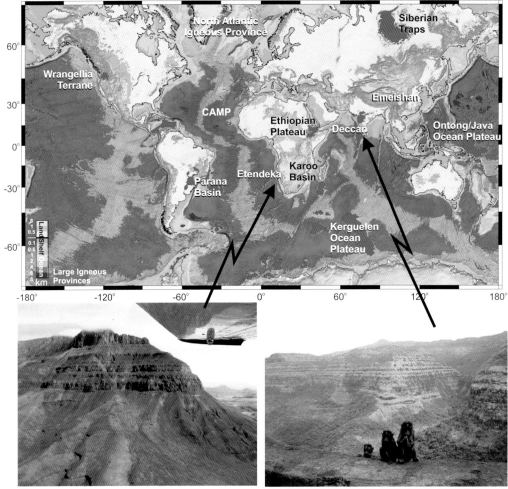

Figure 8.1 Large Igneous Province (LIP) map. The large igneous provinces have some of the thickest sequences of lava flows on the planet. The examples given show stacked lavas from the Etendeka and from the Deccan Traps (Map, Henrick Svensen).

as that of Laki in Iceland, may be pivotal in developing the necessary understanding of the largest flows on Earth. They provide a well-measured example of rapid eruptions where the gases are known, as is their direct effect on the planet.

Volcanic gases and climate

Volcanoes have the potential to change climate, largely due to the fact that they emit a lot of gas. Common gas types released by volcanoes into the atmosphere include water vapour (H_2O), carbon dioxide (CO_2)

Figure 8.2 LIPs and mass extinction correlation (modified from White, R.V. and Saunders, A.D., 2005).

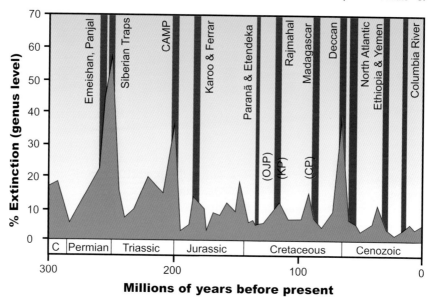

Figure 8.3 Gases from volcanic plumes. Large, explosive volcanic eruptions inject water vapour (H_2O), carbon dioxide (CO_2), sulphur dioxide (SO_2), hydrogen chloride (HCl), hydrogen fluoride (HF), ash and other minor gases, into the stratosphere (USGS).

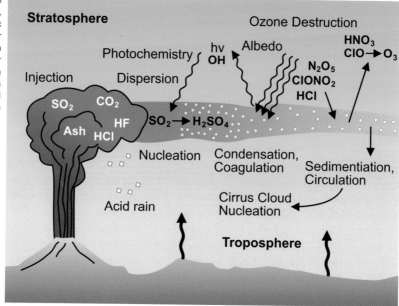

and sulphur dioxide (SO_2). Smaller amounts of other gases are also released, such as carbon monoxide (CO), hydrogen sulphide (H_2S); chlorine and fluorine gases, hydrogen chloride (HCl) and hydrogen fluoride (HF). These gases can affect the atmosphere in a number of ways (Figure 8.3). By far the most common of the volcanically emitted gases that could have a climatic effect are the **carbon gases** and **sulphur gases**. Although the specific nature of the gases emitted depends on the type of eruption and the composition of the magma, all eruptions contain a significant amount of gas. The style of the eruption is also important, as to have the most direct effect the gases need to be sent high into the atmosphere.

The case for volcanic CO_2

The debate about CO_2 in our atmosphere is current. CO_2 has an effect on climate by working as what is known as a **greenhouse gas**. This causes warming as it traps heat radiating from the planet. A certain volume of greenhouse gases is needed in our atmosphere, to maintain a healthy temperature for life to exist, but clearly fluctuations in the CO_2 level in the atmosphere can change that ideal balance, making the temperatures more extreme. The increased amount of **anthropogenic** (man-made) CO_2 is contributing to a global warming event that is measurable. The monitoring of average temperatures, the reduction of the glaciers and ice caps, all point to the impact of the increasing amounts of man-made CO_2 in our atmosphere since the beginning of the industrial age. But what of the contribution of volcanic CO_2? Current estimates of the amount of CO_2 put out by volcanoes are in the range of 100–200 million metric tonnes per year. That seems a large value, but current estimates of the anthropogenic contribution are some 35–40 000 million metric tonnes per year. Thus it would take thousands or more volcanoes to accelerate the current global warming signature. A background level of volcanic CO_2 is being emitted all the time. The largest eruptive episodes, the LIPS, will have had a significantly larger volcanic influence on the atmosphere if their CO_2 level was large.

The case for volcanic sulphur

Gases containing, for example, sulphur dioxide (SO_2) and hydrogen sulphide (H_2S) pose a different potential climatic problem. Instead of a global warming, sulphur in the atmosphere can cause a cooling event. By far the most common sulphur gas is SO_2. If volcanic SO_2 is ejected high into the stratosphere then it can be converted into droplets of sulphuric acid, which condense rapidly to form fine sulphate aerosols. Such aerosols in the high atmosphere reflect the sun's radiation and thus reduce the amount of solar heat reaching the surface. To get SO_2 high in the atmosphere, eruptions need to have high columns or plumes that can deliver the gas to great heights. This is indeed the case with explosive eruptions, such as Pinatubo, *see* below. In many examples of more basaltic systems SO_2 can be in high concentration, yet it may be somewhat rarer for it to be erupted high into the atmosphere. Recent studies have shown that fire-fountaining eruptions in basaltic systems are capable of delivering gases into the upper atmosphere, and these may have formed a key mechanism in some of the flood basalt events that have been implicated in climate change.

Volcanic plumes, particles and haze

Particulate matter, ash and also volcanic gases that are not distributed high into the Earth's atmosphere can produce a variety of problems immediately around a volcano and farther afield. A volcanic haze, sometimes termed **VOG** or volcanic smog, can develop if the air pollution from an erupting volcano is not quickly dissipated by wind. In examples of less explosive, more effusive, eruptions such as those in basaltic type volcanoes, like Hawaii, the VOG is mainly produced by the noxious gases emitted by the eruption. In more explosive examples, the influence of the particles, ash, can become more important. The gases that collect from a volcano can reach deadly levels; for instance, CO_2 gas is more dense than air and will collect in depressions. The eruption of a CO_2-rich plume of gas at Lake Nyos in Cameroon, for example, was caused by the overturn of a volcanic lake that was trapping the gas. The gas moved quietly down the mountain and settled in depressions, killing some 1700 people and thousands of livestock through suffocation. The volcanic plumes of material that rise from explosive eruptions, as in the Iceland volcanic crisis, can eject ash particles into the atmosphere, in this case affecting the flight path of planes. The longer-term effects are limited, as the ash soon falls or is rained out of the sky. Where the volcanic eruption is particularly voluminous and charged with noxious gases, then the volcanic haze can be a real threat to life, as happened during the 1783 Laki eruption.

Modern examples of volcanoes and climate 1 – Laki

A very important eruption in Iceland occurred for an eight month period from June 1783 to February 1784. At a site some 80 km north-east of the location of the recent **Icelandic eruption** at Eyjafjallajökull in Southern Iceland, a series of fissure-fed eruptions occurred in an area near the mountain Laki, termed Lakagígar (craters of Laki). The Laki eruption ejected about 14 km³ of magma, mainly as lava flows and fissure fountains erupting in a relatively short period of time. Most of the outpouring lava occurred in the first five months of the event. The eruption and flow of the lava only formed a direct hazard to local homesteads, but the volcano also emitted vast quantities of noxious gases such as sulphur dioxide and hydrogen fluoride. These gases had a dramatic effect on Iceland, mainly due to the estimated 8 million cubic tonnes of hydrogen fluoride emitted. This caused fluorine poisoning that almost wiped out the island's sheep and dramatically reduced cattle populations. The subsequent famine killed a quarter of the population of Iceland. Laki's effects were felt far from the shores of Iceland. The volcano is estimated to have also emitted some 120 million tonnes of sulphur dioxide gas that drifted as a haze over western Europe, dispersed by the fountaining eruptions that sent the gases high into the atmosphere. The result was one of the hottest summers on record in Europe in 1783; the volcanic smog was so thick that boats were forced to stay in port. North America experienced its coldest winter on record. Laki's far-reaching effects caused crops to fail in Europe, causing famine (e.g.

14 June 1783

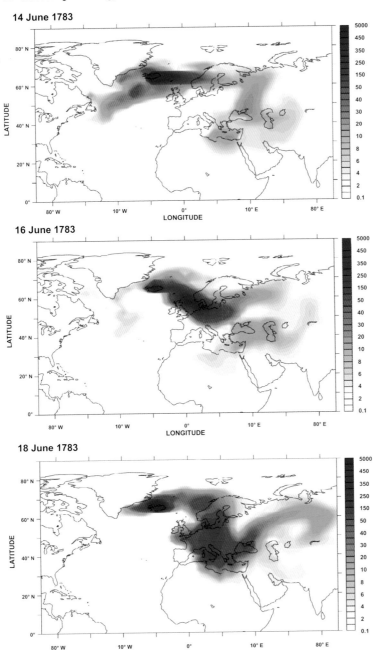

16 June 1783

18 June 1783

Figure 8.4 Modelling the 1783 Laki eruption plume. Numerical models of the distribution of sulphate aerosols in the lowermost (surface) atmosphere for the Laki eruption, Iceland. Model shows every two days from the 14th to the 18th of June 1783, and shows the cloud dispersing over Europe, where it has been cited as causing high mortality (scale on right-hand side in parts per billion) (Vincent Courtillot).

implicated in the Japan famine of this time), and may have contributed to the causes of the French revolution in 1789. In all, Laki is estimated to have played a direct or indirect role in the deaths of several million people, and goes down as one of the major volcanic disasters of modern times.

Using current climate modelling techno-logy, it is possible to create distribution maps that show how the plume from the Laki eruption was dispersed over Europe. These are similar to the models used to map out the effects of the recent Iceland eruption. Figure 8.4 is from a climate modelling system designed by the meteorological laboratory at the Simon Laplace Institute, Paris. Here, the distribution of the volcanic plume from Ice-land's Laki eruption in 1793 is modelled through an atmospheric space divided into 130 000 – 300 km^2 cells in nineteen layers between the ground and altitude up to 40 km high in the atmosphere. Using equations that model atmospheric circulation, the chemical concentration in each cell can be estimated through time. The volcanic products are then added from the position of Iceland and the model runs forward to map out the possible circulation and distribution of the products. In the results, the toxic elements are sometimes blown to Greenland, but are more often sent to Europe. As they reach the lower atmosphere they are leached by rain towards the ground. The dispersion from the model (Figure 8.4) correlates with the maximal extent of the Laki fog cloud estimates, with a ground concentration of ~1000 ppm (parts per million) in Iceland and 50 ppm in western Europe. Thus modelling of past eruptions is becoming more sophis-ticated. But what of the measurements from

modern big eruptions that have happened in the twenty-first century – are the effects more measurable?

Modern examples of volcanoes and climate 2 – Pinatubo

In 1991 a VEI-5–6 eruption took place at **Mt Pinatubo**, the Philippines, erupting some 10 km^3 of magma in an event some ten times greater than that of Mt St Helens in 1980. This was a particularly important eruption, not only for its ferocity but because it allowed direct measurement of the effects of vol-canic gases on the planet. It is a great success story in terms of volcanic prediction and evacuation. On 15th March 1991 volcanic earthquakes around the volcano alerted local villagers. A vigorous monitoring cam-paign then occurred, as earlier geological mapping had revealed very large eruptive events from this volcano in the past and the area was heavily populated. Zones of evacu-ation were defined (depending on volcanic risk) which made up the first 10 km from the volcano, 10–20 km and the remaining 20–40 km. Some 40 000 people lived in zones 1–2 (up to 20 km from the volcano), with over 330 000 overall up to 40 km from the vent. Formal evacuations started on 7th April, with many of the people who lived close to the volcano already moving away some time earlier. By the time of the eruption an esti-mated 60 000 people had left the area 30 km from Pinatubo. The event at Pinatubo occurred with a series of big eruptions during 12th and 13th June, sending an erupt-ion column high up into the atmosphere. It is estimated to have reached some 34 km into the atmosphere during the most violent phase of the eruption (Figure 8.5). The

Figure 8.5 Photo of Pinatubo eruption column. The 12 June 1991 eruption column from Mount Pinatubo taken from the east side of Clark Air Base. U.S. Geological Survey Photograph taken on June 12, 1991, 08:51 hours, by Dave Harlow (USGS).

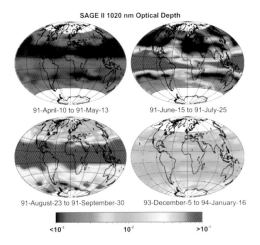

SAGE II 1020 nm Optical Depth

91-April-10 to 91-May-13   91-June-15 to 91-July-25

91-August-23 to 91-September-30   93-December-5 to 94-January-16

<10⁻³   10⁻²   >10⁻¹

Figure 8.6 Pinatubo and climate change. False colour images of aerosol optical depth in the stratosphere, over a period from April 1991 to January 1994. This spans the time before and after the June 1991 eruption of Pinatubo. Red colours show the highest values with dark blue showing the lowest. The Pinatubo/Hudson eruptions increased aerosol optical depth in the stratosphere by a factor of 10 to 100 times normal levels. ('Aerosol optical depth' is a measure of the prevention of light passing through a column of atmosphere by the presence of airborne particles) (NASA).

some 800 deaths suggest that many thousands of lives were saved by the evacuation prior to the eruption.

It is estimated that the Pinatubo eruption ejected nearly 20 million tonnes of SO_2 into the high atmosphere, which lowered the global average temperatures by 0.5–0.6 °C between 1991 and 1993. The monitoring of the aerosol dispersion around the globe is presented in Figure 8.6, and highlights the rapid and

eruption also coincided with a big typhoon, Yunya, which struck north of the volcano, which hampered monitoring and also may have added to casualties. The estimates of effective dissemination of the volcanic gases due to high stratospheric winds. Twenty thousand people, who had lived on the flanks of the volcano, were displaced and the landscape is still scarred by the eruption's aftermath. Many mudflows occurred during the eruption, and after many years are still common in rainy periods.

In August to October 1991 there was also a large eruption at the volcano Mt Hudson in Chile, which ejected 4.3 km^3 of rock plus associated gases in a VEI-5 eruption. These will also have added to the aerosols already in place from Pinatubo, and are likely to have enhanced the levels of stratospheric aerosol (Figure 8.6). The Pinatubo/Hudson event provides direct and compelling evidence that even moderate-sized eruptions can change climatic patterns.

What about a supervolcano eruption?

Volcanoes that have the largest volcanic eruptions – those that exceed 1000 km^3 – have been popularly termed 'supervolcanoes'. Although the individual eruptions that lead to the LIPs can exceed 5000 km^3, perhaps even constituting a new category of '**mega-eruptions**', they are thought to be built up from a number of eruptive vents and fissures, and so are not easily ascribed to any single volcano. Explosive volcanoes with eruptions of VEI-8 or greater are commonly cited as supervolcanoes. There has been much speculation in recent years as to what would happen if a supervolcano were to erupt today, with much of this centring around the Yellowstone caldera in the USA. In the past this volcano has indeed erupted with great violence. At 2.1 Ma, as evidenced by the Huckleberry Ridge tuff ~2500 km^3, an eruption produced a deposit that covered one-third of the North American continent with ash beds several centimetres thick, causing massive disruption nearer the volcano. This and two subsequent eruptions, Mesa Falls, 1.3 Ma and Lava Creek Tuff, 640 000 years ago, created a massive depression in the Earth's crust, a huge volcanic caldera, only really visible from space. Clearly a new super-eruption at Yellowstone would have dramatic effects for life in North America and the world. In a drama-documentary television series, *Supervolcano*, it was suggested that much of the United States of America would be covered by at least one centimetre of ash, with many dead, whilst the global economy would feel the effects of such an eruption. Given our understandings from Pinatubo, it is clear that the climate would be changed, albeit temporarily. The true impact of such a super-eruption is hard to imagine.

Volcanoes and Evolution

As well as their contribution to Earth's atmosphere and moderating climate throughout Earth's history, the role that volcanoes have played and indeed still play in evolution is intriguing. Extinction events caused by the largest outpourings of lavas, the Large Igneous Provinces, no doubt transformed the evolutionary pathway that the planet took at key moments in its long history. Yet it is the creation of new lands that can lead to new habitats, which in turn can lead to new evolutionary adaptations. And with such volcanic influences on evolution it is no wonder that the very theory of evolution sprang from a chain of volcanic islands that highlight this very relationship, the Galápagos.

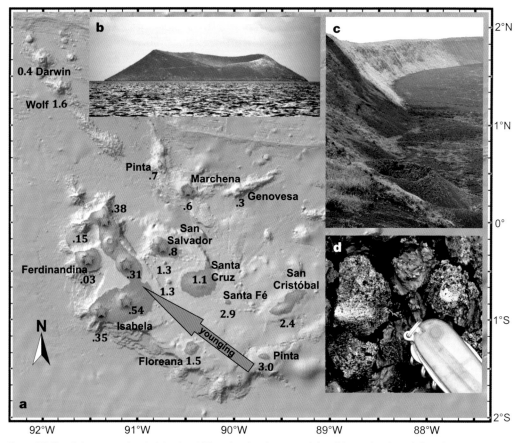

Figure 8.7 The Galapagos volcanic Islands. **a)** Map showing the ages of the oldest volcanic rock found on each island, highlighting the trend to younger ages indicated by the younging arrow (map – Denis Geist). **b)** Volcanic cinder cone island. **c)** Edge of Sierra Negra volcano crater, Isabela Island. D) Fresh scoria from the 2018 Sierra Negra volcano eruption.

Volcanoes that emerge from the ocean floor can create new island worlds for plants and animals to colonise. When Darwin first set foot in the Galápagos, the range of island habitats and vegetation, or lack of it in some cases, was puzzling, as was the wide variety of plant and animal life, separated by relatively short distances between the neighbouring archipelago. Much can be revealed by looking at the evolution of the volcanoes themselves that form the Galápagos island chain. If we explore the relative ages of islands in terms of their oldest volcanic rocks (Figure 8.7), this highlights a remarkable trend that shows that some islands are quite old, and some as young as only a few years. Such variation is driven by the same relationships as the Hawaii island chain (*see* Figure 3.4 in Chapter 3), where a volcanic hotspot (plume) builds up new volcanic lands, while the movement of the plate spreads the lands in time and space. So while one island is old and has been through a number of cycles of colonisation and ecosystem development, new barren volcanic landscapes appear whenever a new island rises from the watery depths, waiting for plants and animals to establish themselves in its new niche environments, and in doing so spark adaptations and evolutionary change.

The rich, diverse and unique species that have developed on the Galápagos are the pages of the book that describes the role of volcanoes and evolution, and the very place where humans discovered how we all evolved from the Earth's fiery beginnings. Amongst the wonders, the giant tortoise is revered as an iconic example of the evolutionary riches of the Galápagos. But for me, it is the cheeky sea iguanas (Galápagos marine iguana), that scurry over the basaltic landscapes, and then dive into the rich blue ocean to feed off the seaweed in the shallows (Figure 8.8), that

Figure 8.8 The Galápagos marine iguana; main photo with an iguana on ropy pāhoehoe lava on Isabela Island; inset photo of an iguana launching itself into the sea in search of a meal of seaweed.

show just what can happen when you get washed up on a deserted volcanic island and have to make a new living of whatever is available to hand.

9   Monitoring volcanoes

'The past is the key to the present', so with all volcanoes a good grasp of how they have behaved in the past is needed to help us understand how they may behave in the future. With active volcanoes the present is very much the key to the future; it's just that recording and interpreting the signals given by modern volcanoes is a developing science. Nevertheless, a catalogue of information is being collected, particularly for the better-monitored volcanoes, where monitoring equipment is permanently housed around the volcano (Figure 9.1). An ever-increasing array of technology is helping to develop an understanding of volcanic activity and its warning signs. Available techniques allow interpretation of signals from deep within

the Earth, from the structures in and around the shallow plumbing of the volcano, from the gases and magmatic behaviour that occur at the vent, and even by monitoring from space. The search is for signals that will allow prediction of when eruptions might happen, how big these eruptions may be, and how long the eruptions will last. These techniques are in their infancy, but with a very well-monitored volcano, there is now better identification of when it is close to eruption. The greatest success at predicting a big eruption was that of the well-monitored Mount Pinatubo. But the signs are not always interpreted correctly, and prediction is difficult either where volcanoes are not monitored or where the volcanic system is very complex. So what are the techniques in use today?

Figure 9.1 Volcano monitoring equipment on top of Stromboli volcano in Italy with volcanic plume in background.

Seismic signals

The most common and, in some cases, most important way in which volcanoes are monitored is by looking at signals coming from beneath the volcano to detect if activity is starting, increasing or changing. This is achieved using **seismometers** (seismographs) which record the shaking of the ground in different directions. The **seismic signals** that are recorded by these instruments can be anything from those produced by a passing lorry or ordnance testing to volcanic and earthquake activity. Networks of seismometers are set up around the volcanoes being monitored, as far away as possible

Figure 9.2 Deploying seismic stations on the side of a volcano. Volcanoes need constant monitoring, and one of the best ways to do this is with seismic stations around the volcano to record any earthquake activity. This example is a seismic station on the flanks of Colima volcano, Mexico. Signals from this and other remote stations are sent to the observatory so that the volcano can be safely monitored from a distance.

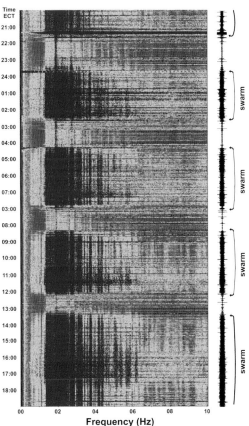

Figure 9.3 Seismic signals from Montserrat. Swarms of earthquakes from 28 November 2009, recorded at a seismic station on Montserrat. The low-frequency volcano seismicity occurs in repeated clusters of a few hours (Montserrat Volcano Observatory).

from areas that have significant extraneous 'noise'. These are used as remote sensors to detect, through vibration, what is happening in the volcano in question. Where detailed networks of seismometers are in place (Figure 9.2), these provide one of the most invaluable types of monitoring and are often the best indications that something is happening in or below the volcano.

The seismic signals are produced by a variety of processes such as magma migration from depth to surface, gas bubble collapse, fracturing, and the flash transformation of groundwater to steam. These processes can vary in their depth within a volcano and produce differing types of signal to be recorded. Volcano-tectonic signals, **VT** or **A-type** signals, are thought to represent the fracturing of wall rocks around magma in dykes and other magma conduits occurring from depth towards shallower parts. Long period signals, **LP** or **B-type**, are low frequency signals found at much more shallow depths in the system and are attributed to bubble collapse, flash steaming and fluid movements. Pulses of air-shock waves during eruptions can produce very short, explosion-type events, and shallow magma movements are often recorded as a volcanic tremor, generally low-frequency **harmonic vibrations**, which often precede eruption and can last from minutes to days. Figure 9.3 highlights an example of seismic signals recorded for the Soufrière Hills, Montserrat, volcano and shows the systematic tremors that cluster as events before eruptions.

Gravity and electromagnetic monitoring

The overall structure of a volcano, even slight changes in the character at local scales, can be studied using gravity. A general survey of the volcanic structure using gravity can provide an understanding of any underlying anomalies such as those that relate to the presence of dense or light bodies beneath the structure. Gravity measurements may also be used to get much closer to understanding what is happening in the magma plumbing before and even during eruptions. To do this, a technique known as **microgravity** provides much more necessary detail. This is a geophysical technique that can measure minute changes in the Earth's gravity. It must be corrected to take account of any ground movements that are occurring at the same time, as these will also have an effect on the overall gravity measurement. Filling of a magma chamber with new magma, or the **vesiculation** of an existing magma reservoir prior to eruption, will change the density of the

surroundings and will change the microgravity. **Electromagnetic fields, EM**, are also used to provide quite a sensitive source of measurement in the shallow parts of a volcanic system, partly due to limitations on the penetration depth that can be recorded. EM signals can be affected by the **electrical resistivity**, which can be changed by fluids, and changes in the geomagnetic field, altered by magma movement and magma type. These measurements, when combined with other remote sensing techniques, can be very powerful tools with which to monitor the shallow volcanic system.

Volcanic gases

The various types of volcanic gases can also be monitored. Changes in the concentration and flux of these can indicate changes in the magma plumbing, can warn of potentially hazardous gas emissions, and can be precursors to volcanic activity. They can be measured directly at the volcano's vent, for instance by measuring **fumaroles**, but this is not a particularly safe methodology. Thus techniques have been developed to measure gas concentrations remotely using radar and light spectroscopy. These include **transform inferred spectroscopy (FTIR)**, **differential optical absorption spectroscopy (DOAS)**, **correlation gas spectrometry (COSPEC)** and **miniaturised UV spec (FLYSPEC)**. The measuring equipment can be mounted a safe distance away from the volcano. In some instances it is placed on vehicles that can be driven through the volcanic plume to take a number of measurements, or at fixed locations where more continuous records of gas emissions can be recorded (Figure 9.4). Examples such as airborne COSPEC can

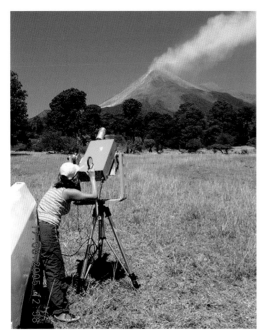

Figure 9.4 Gas monitoring at Colima, Mexico. Correlation gas spectrometry (COSPEC) being measured remotely at Colima Volcano (Julie Roberge).

by ground-based levelling surveys with theodolites, measuring variations between fixed points mapped out in detail and re-surveyed to note changes. It is also done by **tiltmeters**. These take a variety of forms but are essentially devices that record the slope of the ground and how it alters over time. Tiltmeters are a very inexpensive, cost-effective way of determining ground shifts as the volcanic edifice swells prior to eruption. More recently, **Global Positioning Satellite Systems (GPS)** have revolutionised this approach by detecting, with very high precision, the three-dimensional movement of points around a volcano and further afield. GPS networks are being located round the best-monitored volcanoes across the world. Figure 9.5 highlights the GPS movements of stations around Eyjafjallajökull in southern Iceland leading up to the events in March–April 2010. Here, the deformation directly prior to the eruption was picked up by marked deviations of the GPS measurements.

also be used to further expand the potential for analysis. These remote techniques are improving but suffer problems of correlating the signals received to the actual and relative concentration of gases recorded. The difficulties in measuring an active and turbulent gas plume require significant planning for sites. Nonetheless, remote gas measurements are a very powerful and safe way to note continued de-gassing activity at volcanoes.

Tilt and GPS

Ground deformation around the volcano immediately prior to an eruption is a commonly known precursor signal. The standard technique for recording this in the past was

Satellite

Remote sensing techniques using satellites vary widely. Satellites can be used to map out the ground in detail in volcanic areas that are inaccessible, or monitored on the ground, they can be used to determine ground deformation and can record other spectral measurements such as heat emissions. The topography and deformation around a volcano can be measured from satellites, using systems such as **Landsat TM** or **Système Probatoire d'Observation de la Terre, SPOT**. **Digital terrain models**, three-dimensional maps of the topography around volcanoes, can be very informative, particularly where volcanoes are remote or poorly known. Detailed topographic maps and three-dimensional reconstructions can identify where volcanoes have suffered sector collapse and can map out major hazard areas where lahars, debris and pyroclastic flows might be channelled. They can also be used as a base reference from which to build a monitoring campaign around a volcano. A technique known as **INSAR satellite interferometry** is being used increasingly around volcanoes to map out the ground deformation picture before, during and after eruptions. This uses the difference between successive images taken from the same place at different times to see if there have been any ground elevation changes. The concentric interference patterns obtained if satellite data does not match exactly indicates ground movement.

Figure 9.5 GPS movements on Iceland. **a)** Satellite image highlighting location of Eyjafjallajökull and Katla volcanoes (see also Figure 10.9 for location map) and approximate positions of GPS stations indicated (yellow triangles) (image from NASA/JPL). **b)** Displacement profiles from selected GPS stations showing time leading up to the 2010 eruptions. The ground deforms markedly leading up to the March and April eruptions. Grey shading shows the cumulative number of earthquakes with the black shading indicating the daily rate of earthquakes (Sigmundsson *et al.*, 2010).

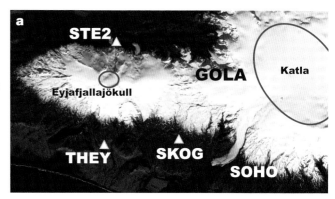

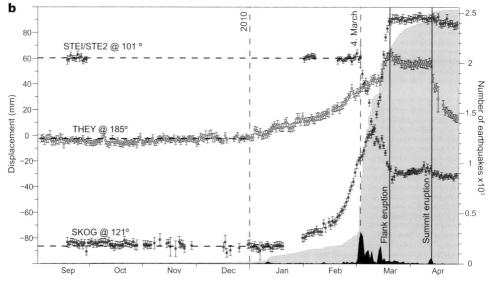

These can be processed to calculate the amount of deformation involved (Figure 9.6). This technique is very powerful, as it can produce deformation maps almost instantly over large areas, in a way almost impossible to achieve using ground-based techniques. There are limitations such as those due to the orbiting times of satellites, the time the satellite needs to orbit and be in the same position to take multiple images of the same place. This is becoming less of an issue as more satellites are deployed, and as satellites can be focused on those areas where monitoring of this kind is required. In remote areas the presence of new crater lakes and eruptions can be established using the information provided about infrared and heat

emissions by, for example, SPOT data. Volcanic plumes from big eruptions can also be monitored by satellites; for example, the **Total Ozone Mapping Spectrometer (TOMS)** and **Modern resolution Imaging Spectroradiometer (MODIS)** have been used to map out sulphate aerosol dispersion and volcanic plumes in the atmosphere.

Thermal

As well as the relatively coarse thermal images that can be obtained from satellite data, it is also possible to use the developing technology of hand-held thermal cameras to measure the heat output and distribution from fumaroles, lava flows and domes. The potential for their application to volcanology is significant. The forward-looking **infrared radiometer (FLIR)** cameras are an example; these can be set to measure temperatures in precise ranges. The cameras take a thermal image of the infrared view at varying pixel resolutions. One issue is calibrating the apparent temperature from the camera to the true temperature, as factors such as air temperature and humidity can affect the results. These difficulties can be overcome, and as with many surveys, the relative change in temperatures between examples is more important than very precise actual temperature measurement. An example of a thermal image of the dome area on top of Colima volcano, Mexico, is provided in Figure 9.7.

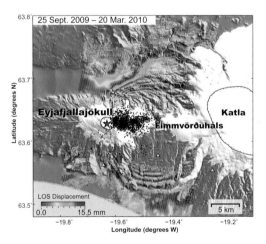

Figure 9.6 Mapping volcanic deformation with Satellites. An example of an INSAR satellite interferometry map from Iceland. The locations of Fimmvörðuháls, Eyjafjallajökull and Katla are indicated. The colour bands correspond to the interference patterns beween satellite maps indicating ground deformation. The black dots show the locations of earthquake epicentres recorded over the same time. Triangles indicate the location of GPS stations (see Figure 9.5) (Adapted from Sigmundsson et al., 2010).

When are volcanoes active, dormant or extinct?

Volcanoes are frequently referred to as being 'dormant' or 'extinct' – it is clear when a volcano is active. What do these terms mean and how are they defined? An 'active' volcano is, somewhat surprisingly, not that easy to

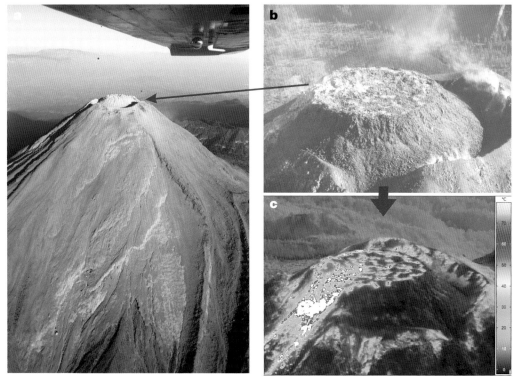

Figure 9.7 Thermal imaging of volcanoes. Example of the thermal imaging of a volcano. **a)** View of Colima stratovolcano from the air; **b)** close up of the crater dome; and **c)** thermal camera image of the dome. The hottest parts of the dome show up as white areas over 80 °C. The dome itself is about 50 m high with a crater width of ~300 m (b & c, Centre of Exchange and Research in Volcanology, Universidad de Colima).

define. For most, an active volcano is one that is currently erupting and/or showing significant degassing and fumarolic activity. However, some definitions, such as those of the Smithsonian Global Volcanology Program, list volcanoes that have erupted in the **Holocene Period**, the last 10 000 years, as active.

A '**dormant**' volcano can be considered as part of an active system, one that has erupted in the last 10 000 years, but has not erupted for some time. Dormant volcanoes range from those that have ceased erupting for several months and show little additional sign of activity, to those that have not erupted for tens, hundreds, or even thousands of years. Clearly this is the most difficult category to define, but it can be important, as people will repopulate and/or move to land around volcanoes when they feel there is no threat of eruption. Also, when people have settled in new areas, they have been shocked when, without apparent warning, a mountain has started to erupt as an active volcano, when there was no perceived hazard before.

An 'extinct' volcano is one that is never expected to erupt again. There is no measurable seismic, fumarolic or other volcanic activity and it has not erupted in historic times. Some definitions suggest that a volcano that has not erupted in the last 10 000 years is extinct. This may appear an open and shut case, but it should be noted that it can be very difficult to be certain that a volcano or volcanic system is extinct. Eruptions have occurred at volcanoes where no known activity had occurred for thousands of years. What shuts a volcanic system off is not understood, nor is what dictates the examples of very long repose times in some systems, perhaps measured in thousands of years. In an ideal world, a detailed picture of each volcanic system could be created, with field mapping, monitoring and a fully collated history of its previous events. This would identify where previous long-repose events have occurred. Such data could then be assessed to determine if it is necessary to have some permanent monitoring in place for volcanoes supposed to be extinct but which have a slight likelihood of erupting again.

The future – building a three-dimensional volcano

Helpful technology is developing very rapidly, and in the geosciences one of the great advances has been the development of true three-dimensional geology. Laser scanning techniques, **LIDAR**, can now produce a virtual outcrop at very high resolution and in remarkably quick time. A virtual outcrop contains all the information about the rock and landscape surfaces in three dimensions. It is akin to being able to take the outcrop home and roam round it on a computer. In volcanology this has a number of advantages. First, where a volcano is unstable, it allows a great deal of information about its structure to be analysed in the safety of the lab. Secondly, repeated three-dimensional analysis shows time-lapse changes in and around the volcano at all scales. The LIDAR can be ground-based or mounted in planes and helicopters. It works by firing millions of laser points at the target, recording the three-dimensional XYZ position of the reflections it receives from the ground. These XYZ point clouds can be coloured according to the ground surface using sequential digital photography or other types of spectral data. The example shown in Figure 9.8 highlights the capture of the first three-dimensional LIDAR image of an active lava lake. The lava lake is Erta Ale, one of the oldest lava lakes in the world, which is located in Ethiopia's Afar desert. To capture the data from this remote volcano, a camel train was required to transport the LIDAR and equipment to the volcano. The final scans of the volcano required the team to abseil down into the heart of the volcano with the equipment. This expedition was filmed for a popular science TV documentary, *The Hottest Place on Earth* (BBC, BBC worldwide, and Discovery Channel) for which the Author was the team geologist. The documentary highlights not only the scanning of the volcano but also the difficulties in working at such remote and poorly monitored volcanic sites. The expedition produced a full three-dimensional model of the inside of the volcanic crater and the surface of the bubbling lava lake: the first survey at this level of detail of this impressive volcano.

An even more exciting development has come with the expansion of the technology in unmanned aerial vehicles (UAVs), known more commonly as drones. These can now be used to fly significant distances and record detailed overlapping images that can be used to map out digital terrain models of the Earth's surface, in a similar way to the LIDAR

Figure 9.8 Virtual volcano; the 3D reconstruction of a lava lake. **a)** Picture of a laser scanner positioned over Erta Ale lava lake in Ethiopia. The scanner takes millions of xyz points to virtually map out its surroundings in 3D. **b)** & **c)** 3D reconstructions of the final virtual model of the lava lake from the scanner information (Jerram and Smith, 2010).

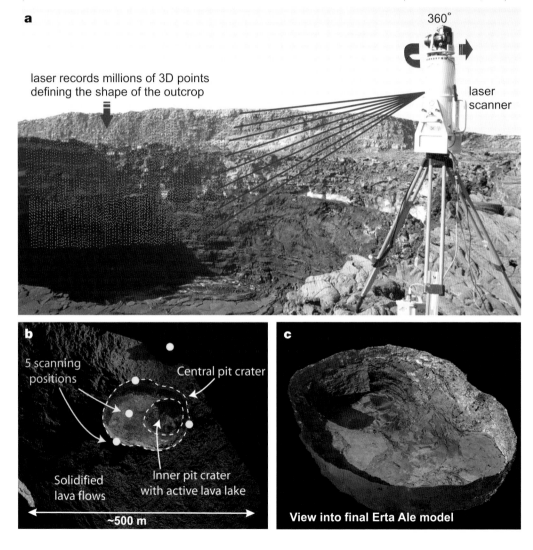

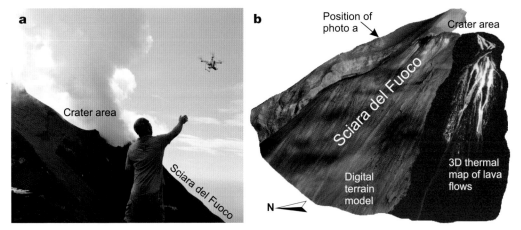

Figure 9.9 3D thermal mapping of a volcano. **a)** Drone launched at the start of a survey with the crater from Stromboli volcano in the background. **b)** 3D digital terrain model of the Sciara del Fuoco slope that cascades down from the crater, with part of the Sciara, where the active lava flow is situated, mapped with a 3D thermal image from a specialist drone with an on-board thermal camera system (images from a joint expedition led by Dougal Jerram and John Howell, Universities of Oslo and Aberdeen).

described above. The value of such technology is that it is light, portable, relatively inexpensive, and can be deployed to more dangerous areas of a volcano. But the excitement and improvements do not stop just there: drones are becoming increasingly more advanced in what they can carry. Figure 9.9 highlights some of the latest technology with a thermal drone used to produce a 3D thermal image of a volcano. This has been tested in detail using the volcano Stromboli as the test bed, with the example shown in Figure 9.9b highlighting the lava flows down the Sciara del Fuoco following the major volcanic event on 3 July 2019. This sort of technology will allow a safer remote, yet very high resolution, monitoring of volcanoes. Exciting future possibilities include the potential to use drone technology to place small high-tech equipment closer to active volcanoes without the risk to life, as well as automated 3D analysis of a volcano, repeatedly mapping any changes through time to provide a 4D monitoring capability. This will allow the creation of interactive **hazard maps** that may be modified as volcanic events dictate. This will also provide a more complete record of how volcanoes change through time.

10 Volcanoes and Man

The Earth has always been volcanically active; its development has been intricately involved with hot rocks. Over history, especially in the modern world, Man has shaped the planet by moulding and developing the environment, building cities, cultivating land and mastering animals, but has not mastered the Earth's most powerful natural processes: earthquakes, erosion, **tsunamis** and volcanoes. In many instances Man lives, or has lived, with volcanoes, and in some key instances, has experienced volcanic forces in more ways than one.

In one sense, Man's relationship with volcanism goes back billions of years to the role that volcanoes played in the development of the atmosphere and of life on the planet. Man lives in proximity to volcanoes in areas beneath the slopes of volcanoes. This uneasy relationship may be illustrated by the volcano **Vesuvius**. The city of Naples and its surrounding settlements lie in the shadow of the volcano (Figure 10.1), and people live there despite the known history of this volcano's eruptions. More remotely, the effects of volcanoes such as Laki have been experienced.

Why Man lives around volcanoes

It may seem odd that people live so close to deadly volcanoes. In some instances the pressure of space on the planet forces us to move to such habitats; where the island itself is built by a volcano, it can be impossible to live anywhere else. It is clear that Man has lived with volcanoes for thousands of years, but why? For the most part the answer is in the soil. Volcanoes provide a constant supply of minerals to the surrounding topography

Figure 10.1 Man living with volcanoes. **a)** View of Naples with Vesuvius towering over the city in the background. **b)** The church at Parícutin in Mexico, half buried by lava from the 1943 eruption.

and fertilise the land that surrounds them. Although a large eruption will first devastate the land immediately around it, the volcanic eruption provides mineral-laden ash to enrich the soil. These ashes weather more readily than solid rock, and the soils that develop from the breakdown of the eruptive products can be very rich and fertile. With regular eruptions through time, the soils are replenished and can be very rich indeed. The lush cultivated slopes around Mount Vesuvius are testament to the virtue of volcanically charged soils. Volcanoes can also be associated with direct **mineralisation** from the sulphur that collects in fumaroles and from the tin, silver, gold and other minerals that can occur in and around volcanic deposits.

The hot steams and springs that are intimately associated with the heating of groundwaters at depth by magma provide another reason why the areas around volcanoes have been populated for centuries. They can provide **geothermal energy**, warmth; and, for some people, hot springs are mythical places where healing and spiritual well-being can be found. A cool beer in a natural hot spring fed from a volcano is a fantastic way to end a day. Man is not alone in this: the Japanese Macaque monkeys, also known as Snow Monkeys, use hot springs to keep warm in the winter months in Yamanouchi, a far northern province, where winter temperatures would not otherwise be tolerable for them (Figure 10.2).

Tourism is playing an increasing role in Man's interaction with volcanoes. Their landscapes are often mountainous, picturesque places teeming with wildlife. They offer the adventure of witnessing an eruption, seeing

Figure 10.2 Snow monkeys of Japan. The author with a *Macaca fuscata* monkey (also known as a Snow Monkey) bathing in Jigokudani Hot Spring in Nagano Prefecture, Japan.

geysers and bubbling pools of mud, or relaxing in volcanic hot springs. For the most part 'volcano tourism' carries no more of a threat to the tourist than any other holiday. Volcanoes are generally quiescent for very long periods of time. There are some that erupt regularly, and some of these, such as Stromboli, are tourist destinations because of this. For the most part we accept the risk of a threatening, large eruption because the lifestyle the volcano offers outweighs its potential threat. The volcanic centre of Yellowstone is an example of this: it is a world-renowned tourist destination, yet it is a supervolcano with the potential for further eruptions. The protracted time between eruptions, the short-term memory of generations and our increasing presence on the planet, removes us from considering what a volcano may do to affect our lives, and the ever-present risk ceases to be a worry. Today, volcanoes near dense centres of population are increasingly well monitored, so the possible threat from an eruption is perceived to be less life-threatening: the risks are perceived to be more to property and livelihood.

Toba – the dawn of Man

Arguably the largest eruption on the planet in the last 2 million years, the **Toba** eruption, occurred approximately 71–72 000 years ago, with a dating error of about 4000 years. This was a VEI-8 eruption which involved some 2800 km³ of material. It is estimated that the eruption covered some 20 000 km² with deposits as thick as 600 m at points nearest the volcanic vents. The event is recorded in ice cores from Greenland, where reduced levels of organic carbon sequestration can be observed. The **volcanic winter** that would have started after the eruption would have lasted several years, resulting in a global temperature reduction of 3–3.5 °C. The effects in the SE Asia region would have been catastrophic (Figure 10.3). Was the volcano potentially more deadly in its aftermath than can be imagined? Could the eruption have nearly killed off Man altogether?

Some theories propose that the Toba eruption came close to stopping the evolution of Man in its tracks. Using evidence from **mitochondrial DNA** it has been suggested that Man's evolution went through a genetic 'bottleneck' at a time contemporaneous with the Toba eruption. The suggested bottleneck is evidenced, it is suggested, by a reduction in Man's genetic diversity well below what would be expected of a species of our age of development. The Toba eruption, it is argued, caused this reduction in genetic diversity as Man's numbers reduced to dangerously low levels due to the climatic and biological effects that followed the eruption. If this is the case, then the world's human population may have been reduced to only a few tens of thousands of individuals by the Toba eruption. Uncertainties in the dating of the eruption and the dating of the DNA bottleneck cause problems for the theory. There are also uncertainties in the estimates for the impact of the eruption and how much of a decline in Man's numbers can be attributed

Figure 10.3 The Toba eruption. **a)** Landsat satellite image highlighting location of Lake Toba (NASA). **b)** Summary geological map showing extent of preserved tuff and major caldera and structural faults.

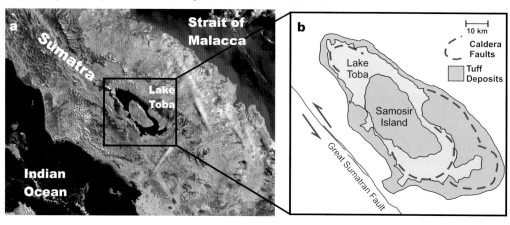

to it. Even so, Toba was a massive eruption, and its links to a dramatically changed climate with a consequent impact on population levels make for a compelling argument.

Vesuvius – destruction of Pompeii and Herculaneum

The most famous eruption ever recorded in western Europe, arguably the world, was the eruption on the 24th August AD 79 of Vesuvius, an explosive eruption that rocked the Bay of Naples. The eruption provides a remarkable catalogue of information. The subject of volcanology may be said to have begun with the vivid descriptions of the volcanic activity that occurred. The power of the eruption destroyed two towns and fossilised a whole community. It provides an eerie window into the everyday lives of the Romans then living in the area and how their lives came to an abrupt end due to the volcano.

One of the oldest detailed descriptions of a volcanic eruption was written by a scholar, Pliny the Younger, who watched the whole devastating episode. Importantly, with his record and the deposits from the eruption itself, it is possible to unravel details of the event in a manner rarely realised from similar eruptions in the distant past. Two towns, Pompeii and Herculaneum, were buried during the eruption, yet the tales from each town, separated by only a few kilometres, provide separate and differing accounts of the eruption's dynamics and lead to a better understanding of the eruption.

The slow death of Pompeii

Pompeii was a lively market town with many holiday villas, public baths, an amphitheatre, an aqueduct system and a town brothel. It was at a crossroads of routes between a number of towns and provided a

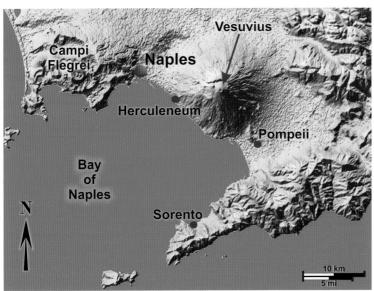

Figure 10.4 Map of the Bay of Naples. Bay of Naples highlighting positions of Herculaneum and Pompeii in relation to Vesuvius (Martin C. Doege, based on NASA SRTM3 data).

safe harbour, inland from the sea. Lying some ten or so kilometres from the present Vesuvius structure, it would have had very picturesque views of the mountain and the Bay of Naples (Figure 10.4). About 20 000 people lived at Pompeii before the eruption and estimates suggest that about 2000 people were killed. This suggests that there was sufficient warning of the eruption to allow the evacuation of many of Pompeii's inhabitants before it erupted (Figure 10.5). It is likely that many people died when the later stages of the eruption overcame the town – with those dying being people who refused to go or could not easily leave: people returning after the initial evacuation to retrieve belongings, and even, possibly, looters. It is clear from Pliny's descriptions that the eruption continued for some time, and people, including his uncle, Pliny the Elder, were working at rescue and evacuation when the volcano finally took their lives.

The preservation of the bodies at Pompeii is in itself interesting. During early excavations it was noticed that cavities in the volcanic ash were being discovered. The remains of bones, full human or animal skeletons, were found in each of the cavities. During the 1860s, Giuseppe Fiorelli, the Director of Excavations, developed a method of investigating the nature of the cavities by filling them with plaster of Paris. The soft ash deposit surrounding each cavity was then removed once the plaster set, to reveal perfect moulds of the bodies of the people and animals that died (Figure 10.6). These body moulds tell the story of the final moments of life in Pompeii, a harrowing tale of suffocation under a densifying air, heavy with ash from the eruption. While this was

Figure 10.5 Eruption of Vesuvius reconstructed. Computer reconstruction of the last days of Pompeii as it is enveloped by the eruption of Vesuvius (Crew Creative Ltd).

happening in Pompeii, a different catalogue of events was unfolding at the port town of Herculaneum, some 20 km away.

The rapid death of Herculaneum

Herculaneum was a seaport significantly closer to Vesuvius than Pompeii, at about 7 km. Thus it would be logical to expect that more fatalities, relative to the town's population, would be found in Herculaneum than at Pompeii. The initial excavations of the site were very puzzling. Due to its proximity to the volcano the town was buried under more substantial volcanic deposits. This made excavation a little harder, and the

Figure 10.6 Body casts at Pompeii. The casts are made by plaster of paris being poured into body cavities discovered while excavating the ruins at Pompeii. They show harrowing details of people and animals struggling with their final breath as they are overcome by the volcanic cloud when the eruption of Vesuvius engulfed the town.

site itself was only fully realised somewhat later than that at Pompeii. However, the thicker deposits meant that in places, the town was preserved more pristinely than Pompeii. The puzzling aspect of the excavations was the lack of bodies when more might be expected, as the site of Herculaneum is close to Vesuvius (Figure 10.4). It appeared that the town was deserted when it was overcome by the cloud of volcanic ash and rocks that buried it. Initially it was believed that the inhabitants had time to escape, since Herculaneum was a port where quick access to boats could have aided escape. However, when the diggings reached the port area the stark reality was revealed in the boathouses there, where 300 or more bodies were discovered. It is now suggested that the people had gone here to shelter from the eruption, unaware that the town was about to be buried under tens of metres of volcanic rubble.

Blackened
bone end

Figure 10.7 Bones at Herculaneum. Bone beds found in the boat houses at the site of the old port in Herculaneum. The discovery of these bodies indicated that the people were taking shelter in the boat houses, maybe waiting for rescue from boats in the Bay of Naples. The bones show blackened/burned parts (indicated in photo **b)**, which have been used to show that the people were taken by surprise by a very hot pyroclastic flow, instantly entombing them in their boat-house graves.

A striking feature of this discovery is that there are no body moulds such as those found in Pompeii; just perfectly preserved skeletons, as if they had been stripped of life where they stood or sheltered (Figure 10.7). So what caused these differences? Close inspection of the bones at Herculaneum reveals that many have blackened coloration around the joints, where the body's skin is thin. Experiments to

try to reproduce this blackening show that exposure, for short periods, to flash heating at high temperatures can produce blackening on those parts of the bone where there is little flesh. Where muscle and other tissue protect the bone, the blackening is less apparent (Figure 10.7). The pyroclastic density currents that barrelled down the mountain and buried Herculaneum were not only deadly in terms of their ash and blocks of rock; they also reached a temperature in excess of 500 °C, capable of stripping the body of its flesh and burning the bones underneath. The rock and ash acted as geological cement, to preserve this process for us to reveal in the modern excavations. This is the first known preservation of human remains by hot pyroclastic rocks. In Herculaneum the people had fore-warning; they will have seen the mushroom cloud form that Pliny reported. Many people may have escaped, but those that sought the shelter of the boathouses stood no chance as the hot pyroclastic flows sealed their fate.

The deposits at the foot of Vesuvius remain some of the best-studied examples of the historic effects of volcanoes on Man. Not only do the towns of Herculaneum and Pompeii, and their deposits, provide a record of the volcanic eruption itself; they are regarded as world class archaeological sites. The added detail of the accounts of the eruption by Pliny, contained in two letters that he wrote to the Roman historian, Tacitus, make it a unique volcanological dataset which is still providing information as the excavations continue. The function of volcanic deposits when they blanket and fossilise the environment is to provide a unique mechanism to preserve things, from wood and buildings to people and animals.

Mt St Helens

The Cascades mountain range in the USA is home to a number of volcanoes, and one of the most active is Mt St Helens. This volcano in Oregon State has erupted more frequently than others in the range during recent times (Figure 4.8, Chapter 4), but few were prepared for what happened in 1980. Mt St Helens is very well monitored, and signs that it was about to erupt were first noticed at 43 seconds after 3:47pm on Thursday 20th March 1980. Such precision in timing testifies to the level of monitoring. Seismic tremors from beneath the mountain were recorded by stations surrounding the volcano. Some small summit eruptions followed, suggesting that the volcano had reawakened after 123 years of quiet. The monitoring of the volcano contin-ued and, by late April, the ground around the northern flank of the volcano began to swell. The bulge was caused by fresh magma being pumped into the shallow structure of the mountain. At 8:32 on 18th May 1980 an earth-quake of magnitude 5 shook the mountain and caused a landslide on the northern face of the mountain, which had been weakened by the swelling. As the weight of rock covering the bulge slipped away the fresh, gas-filled magma was released like the cork from a champagne bottle and the eruption started in earnest. This was not the normal eruption that volcanologists were expecting, as the entire side of the volcano had fallen away. The collapse and eruption sent pyroclastic flows, rock avalanches and mudflows towards the north. A gigantic plume of hot ash towered into the sky with a column height of 19 km. The ash plume was then to spread over large parts of mainland USA. The overall size of the eruption was ranked at VEI-5.

What was so special about the Mt St Helens' 1980 eruption was that it was the first time a **lateral blast** from a volcano had been witnessed. This volcano has shown that a stratovolcano can collapse and erupt with a large sideways force. The result was a crater 2–3 km wide that was opened to the north. The mountain was reduced in height by 400 m (Figure 10.8). The volcano continued to be active, and soon a resurgent dome of new magma was beginning the process of rebuilding the mountain. Fifty-seven people were killed as a direct result of the eruption, mostly due to the unexpected lateral blast; a vertical eruption had been expected. This number would have been far greater had not the geological and geophysical monitoring provided sufficient warning for exclusion zones to be put in place around the volcano.

Following the Mt St Helens' 1980 erupt-ion, a number of stratovolcanoes have had their deposits re-interpreted with better knowledge of such lateral blast phenomena, known as **sector collapse** or **debris ava-lanche**. This may not seem important, since these deposits have already been erupted, but their very identification allows for re-evaluation of the potential hazard posed by the volcanoes surveyed. New volcano hazard maps and behavioural models take into account the possibility of sector collapse. This requires not only a good understanding of the previous deposits from old eruptions, but also an understanding of the structure of the volcano, so as to predict areas of weak-ness. The new Mt St Helens has nearly re-emerged; it is instructive how quickly such violent volcanic events can be hidden and covered up by the volcano itself and by weathering and erosion.

Figure 10.8 Mt St Helens before and after the eruption. **a)** Before the devastating 18 May 1980 eruption (US Forest Service). **b)** After the eruption, photograph taken on 19 May 1982 (U.S. Geological Survey). **c)** View looking north from the top of Mt St Helens with new dome growing in the crater, Sept 2001.

The billion dollar volcano

The so-called 'Icelandic volcanic crisis' of 2010 reputedly cost the global economy several billion dollars. It created a situation unprecedented by other natural disasters by directly affecting the lifestyle of many millions of people, without causing death or directly threatening their lives. The volcano's site was remote, yet its direct influence on our lives was both very pronounced and also very rapid. The eruption caused the airspace over most of Europe to be closed for days. This disrupted world travel and the skies over cities such as London were quieter than

people could remember for decades. It was reported that residents heard birdsong in the skies around Heathrow Airport, London. This was caused by the eruption of Eyjafjallajökull; the 'Billion Dollar Volcano'.

Leading up to the eruption there were plenty of signs that things were going to happen. Iceland has a very good network of remote sensing equipment that monitors its many volcanoes, and the Nordic Volcanological Centre (at the Institute of Earth Sciences, University of Iceland) and Iceland's meteorological service, keep a watchful eye. Seismic activity was on the increase from

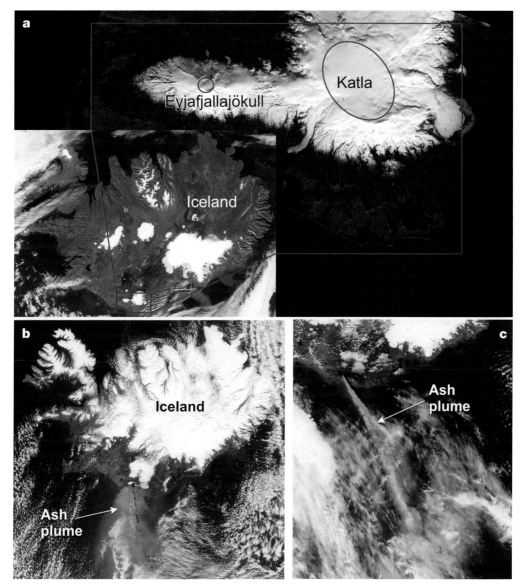

Figure 10.9 Satellite images of the Iceland and the 2010 plume. **a)**The location of Eyjafjallajökull and Katla on Iceland. Satellite image of Iceland highlighting location of Eyjafjallajökull and Katla volcanoes (box) with enlarged region showing the ice-capped volcano centres in the south of Iceland. (NASA/GSFC and NASA/JPL) **b)** NASA's Aqua satellite visible image of the ash plume from the Eyjafjallajökull volcano 17 April 2010 (NASA's MODIS Rapid Response Team). **c)** NASA's Aqua satellite image over Iceland's Eyjafjallajökull Volcano on 10 May 2010 (NASA Goddard/MODIS Rapid Response Team).

Figure 10.10 Eruptions from the Iceland volcanic crisis: the different stages of eruption during the Iceland volcanic crisis in 2010. **a)** Fissure-fed fire-fountaining during the early stages of eruption located at Fimmvörðuháls, on the saddle between Eyjafjallajökull and Katla. **b)** Explosive eruptions at Eyjafjallajökull itself as the eruption centre shifts to the main volcano and interacts with the ice cap (Jón Viðar Sigurðsson).

December 2009, highly increased activity occurring from around the 26th of February, and the ground was starting to move, evidenced by **GPS stations** around the volcano. The first eruption is thought to have begun on 20th March 2010 about 8 km east of Eyjafjallajökull's crater, on Fimmvörðuháls, situated on the saddle between Eyjafjallajökull and Katla volcanoes (Figure 10.9). This early eruptive phase produced fire fountains, small explosions and some lava. Importantly, at Fimmvörðuháls there was no ice cover, and so the magma was erupting directly out of the ground (Figure 10.10a). After a short pause, eruptions started on 14th April 2010 at the main Eyjafjallajökull volcano, under a small ice cap. This change in the eruptive centre resulted in much more explosive volcanism. The new magma batch had interrupted the chamber beneath Eyjafjallajökull itself. A larger batch of more evolved explosive magma was being erupted, and its dynamic interaction with ice, flashing to steam, fuelled the VEI-3–4 eruption. An estimated 250 million cubic metres (¼ km³) of material was ejected with a plume height of up to 9 km, 30 000 ft (Figure 10.10b).

This change in the eruptive activity delivered the ash into high wind systems blowing towards mainland Europe. The result was an ash plume that was directed into European airspace, shutting down all flights over much of the area for six days. Satellite images from the time highlight the way in which the plume was being delivered to Europe (Figure 10.9b & c). Before the eruption there were civil aviation authority regulations regarding flights through volcanic ash and, although the ash was not in very high concentrations in all parts of the affected airspace, there was zero tolerance of flying through ash. In the past, aircraft have dropped out of the sky as their engines have failed, clogged by volcanic ash, and have only managed to restart their engines, close to the ground, in time to avert disaster. The zero tolerance limit was being tested as the volcanic eruption was lasting for days and there seemed no foreseeable end to the disruption. A previous eruption, in 1821, had lasted over a year. Following urgent tests and with mounting public pressure, the flight rules were changed and established a new tolerance limit of 2000 µg/m³; anything more than 2000 µg/m³ is still a no-fly zone. At between 2000 and 200 µg/m³ planes are required to take extra precautions; below 200 µg/m³ no threat is recognised. For comparison, the EU recommendation for limits on particulates in air pollution is only 20 µg/m³ (yearly average). A tolerance of 4 000 µg/m³ was also accepted for certain areas for limited times in UK airspace, one of the busiest on the planet, and that which was being most affected by the plume. Had these limits been in place at the start of the crisis, very limited areas of airspace would have been closed and for shorter time periods. The threat still remains that in favourable wind conditions eruptions from Iceland and other volcanoes around the world will shut down airspace.

Watching the birth of a volcano

The largest eruption on Iceland since the Laki flow in 1783 (the 2014–2015 Holuhraun fissure eruption), and one that was watched in the digital age, occurred just north of the Vatnajökull glacier between 31st August 2014 and 27th February 2015 (Figure 10.11). The first signs of activity were recorded as earthquake

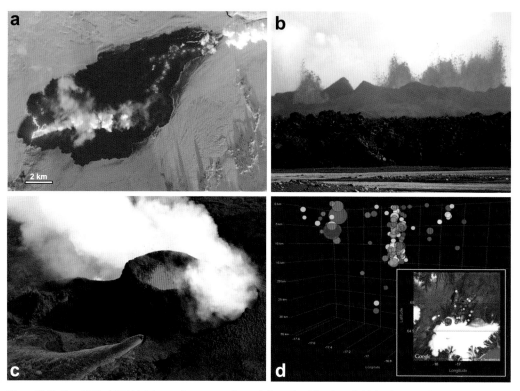

Figure 10.11 The 2014–-2015 Holuhraun fissure eruption. **a)** Growth of the lava field by January 2015, imaged using Landsat 8 (NASA). **b)** Fire fountain in background with advancing lava front in foreground. **c)** Lava breakout flow from the side of 'Heiturpottur' the 'hot pot', officially named Suðri (southern) crater after the eruption, (photos b&c taken on 10 September 2014 by a team from University of Oslo, VBPR, DougalEARTH and the University of Iceland – Dougal Jerram, Sverre Planke, Morgan Jones, John Millett and Sigurður Gíslason). **d)** 3D Seismic monitoring of the activity around Barðarbunga and Holuhraun (cr. 29/08/2014 – larger circles are magnitude greater than 4, Blue colour over a day ago, Red most recent) (images courtesy of the web resource '3dBulge' created by Bæring Steinþórsson using data provided by the Iceland Met Office).

tremors from magma movement underground as magma was intruding its way progressively northwards from the nearby Barðarbunga volcanic centre. On 29th August 2014 there was a shortlived minor eruption lasting just a few hours, but the main eruption phase started on the morning of the 31st August. The eruption was characterised by lava fountaining from fissure eruptions, and an extensive basaltic lava flow that measured some 1.6 km³ in volume when it finished in February 2015, marking a period of approximately six months of continuous activity. At its height the maximum fissure length was about 1.5 km and the final flow covered an area of some 84 km².

Here, in many ways, we were able to record the 'birth of a volcano', due to the extensive monitoring system that is in place in Iceland, and the Holuhraun fissure eruption was remarkable in many ways. There was significant seismic activity from magma movement leading up to the eruption form around the 16th of August up to and during the eruption itself, providing a day to day account of the heartbeat of the volcano, and which marked the pathway of the magma towards the surface. As the tremors from magma movement underground continued, the monitoring at and around Barðarbunga meant that the 3D location of the earthquakes could be visualised (Figure 10.11d) Therefore, as the magma migrated through the crust it was observable virtually by its earthquake epicentre trail. Even after the fissure had started to erupt, the dynamics of the system were constantly under surveillance, with earthquakes and ground movements being recorded. The system associated with the Holuhraun eruption was linked back to Barðarbunga volcano and a period of caldera collapse. This highlighted the link between caldera collapse events and flank fissure eruptions, whereas a more classical view of caldera collapse was associated with explosive eruptions in and around the caldera edges (*see* 'Giant holes in the ground', Chapter 4).

Creation and destruction of land

Volcanoes present an unstoppable force that has the power to be both destructive and creative. They can remove whole mountains (e.g. Mt St Helens eruption 1980) and create new lands (Surtsey, Iceland 1963). The plethora of volcanic islands and archipelagos around the world are testament to the creative and destructive nature of volcanic systems, with new islands being created, volcanic collapse events causing destructions and even tsunamis, and the eventual erosion and demise of these new lands with the passing of time and of plate tectonic motions (cf. Hawaii volcanic chain, Figure 3.4). With the recent ascent of Man and his colonisation of volcanic worlds, the battle for Man's stake in new lands can be a precarious one.

The recent 2018 lower Puna eruption sequence at Kīlauea Volcano and its Lower East Rift Zone (LERZ) on Hawaii provide a window into the hazards faced and the battle for ground that man can have with an active volcanic system (Figure 10.12). The eruption and associated activity started in 30th April 2018 with collapse at the long-term Pu'u 'Ō'ō eruptive vent and magma movement down the rift, with the deflation of Kīlauea's summit area and Halema'uma'u lava lake level fluctuations on 1st May. Cracks first appeared around Leilani Estates residential area along the LERZ on the 2nd of May and flank eruptions start on 3rd May, in a sequence of fissure eruptions and lava flows that would continue into early September and see the destruction of houses, roads and communities, as well as the creation of new land along a ~7 km of stretch of the coastline (Figure 10.12). In total, lava flows covering an area of some 35.5 km^2 (13.7 m^2) were erupted, showing how eruptions from one part of the island can have a marked effect over a much larger footprint. In the case of these fissure eruptions along the LERZ, the fissure labelled 'fissure 8' resulted in flows with the most coverage, with lava reaching the ocean from fissure 8 at Kapoho Bay (*see* Figure 10.12). The event saw the destruction of 716 dwellings,

Figure 10.12 Kīlauea 2018 eruption. a & **b)** Maps highlighting the progress of the eruption and aerial extent of the flows from May to August 2018. **c)** At 07:45 a.m. local time on 5 May 2018, lava from a fissure slowly advanced to the northeast on Hoʻokupu Street in Leilani Estates subdivision on Kīlauea Volcano's lower East Rift Zone. **d)** Fast-moving lava flows erupted from the fissure 8 cone (lower right), 29 July 2018 (USGS).

~30 miles of roads covered with lava, and a total of 875 new acres of land being created (statistics from USGS).

In addition to the fissure eruptions and lava flows, activity back at the Kīlauea Caldera was similarly dramatic and earth-moving. With the continuing eruptions along the fissure system the caldera itself was deflating. This is a similar caldera collapse event to that seen in the Holuhraun/Barðabunga eruption in Iceland in 2014 (*see* also 'Big holes in the Ground', Chapter 4). The 'before' and 'after' images in Figure 10.13 show the areal extent of this caldera collapse between June and August 2018. The cross-sectional view of Kīlauea's summit topography (Figure 10.13c), provides a stark visual image of the extent of the drop in the caldera elevation, reaching a maximum of nearly half a kilometre at its deepest drop.

Volcanic tsunamis
Another dramatic result of the relationship of volcanoes with the sea/ocean is that of volcanically induced tsunamis. Volcanic tsunamis are giant waves caused by mass movements into water bodies, and can result from volcano island collapse, eruption into

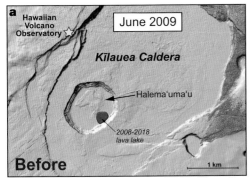

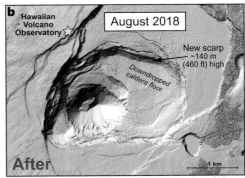

The volume of Halema'uma'u was around 54-60 million cubic meters (70-78 million cubic yards) prior to the 2018 events.

The volume of the inner collapse crater is now about 885 million cubic meters (1.2 billion cubic yards). Subsidence of the adjacent caldera floor created the new scarp (arrow).

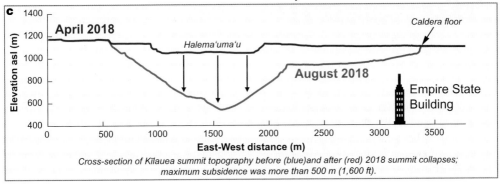

Cross-section of Kīlauea summit topography before (blue)and after (red) 2018 summit collapses; maximum subsidence was more than 500 m (1,600 ft).

Figure 10.13 Collapse of the Kīlauea Caldera during the fissure eruption in 2018. **a)** Topographic image prior to the eruption in June 2009 showing Halema'uma'u crater and the location of the lava lake. **b)** New morphology of the crater after the collapse in August 2018. C) Cross section before (blue) and after (red) the collapse at Kīlauea, with maximum subsidence over 500 m (1,600 ft) with Empire State Building for scale (USGS).

water and tectonic movement from volcanic activity. The most infamous volcanic tsunamis were caused by the eruption and collapse of Krakatau (Krakatoa) in 1883. This violent eruptive episode took place between 26th and 27th August, and saw eruptions ranging from submarine Surtseyan to plinian, with a gigantic ultra-plinian VEI-6 eruption and associated flank failure and collapse of the caldera, which resulted in a number of tsunamis, the most destructive being generated

around 10 am on 27th August. The aftermath of the eruption resulted in around 70% of the island of Krakatau and its surrounding archipelago being destroyed, and resulted in more than 36 000 fatalities, most of which were caused by the volcanic tsunamis. Some decades after this fateful event, a new volcanic island (Child of Krakatau) was constructed within the 1883 caldera with recorded frequent eruptions since 1927. Anak Krakatau, although the smaller child of its

Figure 10.14 Aftermath image of damage following the Anak Krakatau Tsunamis 2018 (Indonesian National Armed Forces).

larger volcano, would see history repeat itself in another volcanic tsunami event.

Move forward to 22 December 2018, and an eruption at Anak Krakatau and its associated cone collapse caused a deadly tsunami, resulting in some 437 deaths and thousands being injured or displaced along the surrounding coastline. With waves up to five metres in height hitting the adjacent coasts, and at a time when people were going about their business with little warning, the volcanic tsunamis crashed into the land. In the digital age, the events were recorded due to a concert that was taking place at the time, and the destructive nature of the waves could be seen in the aftermath (Figure 10.14). The Krakatau volcanic system, and its dynamic association with the surrounding waters, mean that it will be a volcanic tsunami hazard into the future. Warning systems for tsunamis can be put into place in order to help mitigate the deadly, if not the destructive, events of a tsunami. One such system has been installed on the volcanic island of Stromboli following a landslide event that occurred on 30th December 2002, when two closely spaced landslides triggered tsunamis. A total estimate of 5 600 000 m³ of rock detached from the Sciara del Fuoco, the main volcanic flank of Stromboli (*see* Figure 9.9), and cascaded into the sea below. Two tsunamis resulted, causing damage to buildings and boats, and thankfully only causing minor injuries to people, as this had occurred in a quiet time on the island. Since this event a tsunami warning system (with a buoy tiltmeter in the sea) has been installed and signs around the town are in place to direct people to higher ground in the event of an alarm. In the case of tsunamis, sometimes only a short warning of several seconds can be enough to get to higher ground and potential safety.

Volcano tourism and recent events

Volcanoes and their environments are beautiful. They provide an array of fantastical landscapes and in many ways are a link to the primordial past of Earth and direct evidence of our dynamic breathing planet. It is no wonder, then, that volcanoes have drawn Man's attention for many thousands of years. In the more recent past, the role of volcanoes in tourism has become increasingly important. This can be wide and far reaching, from regions whose landscape is almost entirely volcanic (e.g. Canary Islands, the Azores, Galápagos, Hawaiian Islands, Réunion), mountain ranges with active volcanic systems (e.g. Massif Central, Cascades, Andes), to the individual volcanoes themselves. Classic volcanoes such as Vesuvius, Etna, Stromboli, Santorini, Yellowstone, amongst many others are the sites of volcano tourism, where people go to see hydrothermal pools, bubbling mudpots, dramatic craters and even the eruptions themselves.

However, volcanic terrains that are associated with dormant and active systems carry associated risks. Even extinct volcanic landscapes can be susceptible to hazards such as landslides due to their relatively juvenile and un-vegetated landforms. Some recent eruptive events at popular volcanoes have touched on the delicate nature of volcanoes and tourism. On 27 September 2014, a surprise eruption at Mount Ontake in Japan killed 63 people. Mount Ontake is the second highest volcano in Japan and for thousands of years has drawn worshippers and pilgrims to its slopes and the huts along the pilgrim path. Many people were on the volcano on the day of the eruption, which was caused by a shallow phreatic eruption and associated pyroclastic density currents. The eruption was sudden and did not seem to have any clear precursory surface phenomena such as rumblings or increased gas activity. Recently the Mountain has reopened to tourist activity with a 1 km exclusion zone around the crater. A similar phreatic eruption with little warning occurred at Whakaari/White Island volcano on 9 December 2019, killing 21 people and injuring a further 26. The area around the crater of White Island has been a popular place to visit for many years with its hydrothermal system and fumaroles and its wonderfully stark and colourful landscape (Figure 10.15). Similar to the Ontake eruption, the nature of the shallow interaction of the volcano with subterranean waters, causing the shallow phreatic eruptions, provided few 'normal' precursor indicators.

In other examples, the relative timings of larger volcanic events in systems that are generally more predictable can play a role in the outcome of the eruption from Man's perspective. The island of Stromboli is one of the world's most iconic volcanic islands and is a big draw for tourists who come to visit Stromboli, the surrounding Aeolian Islands and the volcano Etna on Sicily. It is common for tourist groups to visit the summit of the volcano, which has 'strombolian' activity (named after the volcano), which can occur every 15 minutes or so. However, even a relatively predictable and stable system such as Stromboli occasionally has larger eruptions termed 'paroxysms', which carry more risks. On the 3rd of July 2019, Stromboli experienced a large paroxysm event, probably due to a large magma bubble-driven eruption, which showered the upper slopes with hot incandescent fragments (causing some

Figure 10.15 Group of tourists trekking around the crater at Whakaari/White Island volcano prior to the 2019 eruption (© Philippe Gauchot/Shutterstock).

ground fires), raining ash on the villages, and resulting in an eruption column rising kilometres into the sky (Figure 10.16). Following the paroxysm, a lava flow system developed on the flank, running down the the Sciara del Fuoco toward the sea (Figure 10.15c). Unfortunately one tourist was killed during the eruption, but due to the timing of the event in the afternoon, with most tourists climbing up for early evening, it meant that few people were on the mountain and any resultant fatalities limited. A further explosive event occurred on the 28th of August 2019, when a pyroclastic density current was produced, which was further indication

of a more active phase of the volcano during this time.

Within all of these volcanic systems, the key to understanding and mitigating their risks is in the detailed monitoring of the volcanoes, as well as developing an understanding of the volcanoes' behaviour through time. As we get more eruptive events occurring where detailed monitoring systems are in place, the detail held within such data may help us to unravel the 'key' signals/behaviours that act as precursor warnings to the eruptions. People will still be drawn to volcanic landscapes, and to that extent volcano tourism is still very much a growing business, which

Figure 10.16 The 2019 Stromboli paroxysm. **a)** View of the eruption column from a beach on Stromboli (Shutterstock image). **b)** Ash from the eruption cloud as well as smoke from the resulting ground fires on the flank of Stromboli, as viewed from residence on Stromboli (Sara Sorrentino). **c)** Strombolian eruption with lava flows making their way down the flank of Sciara del Fuoco on 24 July 2019 (John Howell).

needs to adapt to the demands of both Man and volcano alike, such that we can enjoy volcanoes with as little risk as possible.

Summary

The Earth is an active planet where volcanism is a powerful force beyond Man's control, and a force that has played a direct role in the development of life on Earth. As we increase occupation of our planet, there is more and closer contact with moderate-sized volcanoes and their eruptions. Unravelling the planet's geological secrets reveals some of its past mega-eruptions, and where and when these occurred. A message we can draw from this is that the biggest eruptions that the Earth can produce have not been witnessed, and, whilst it is not clear when they may happen, it is certain that they will happen again. Looking to the future, our modern lifestyles are also susceptible to the remote effects of volcanoes, as has been demonstrated by the effects of the plume from Eyjafjallajökull. People who live in direct contact with volcanoes know they provide valuable resources and wealth, but also need to respect their power and the risks they present. A volcanic eruption is one of the wonders of the world; it can be witnessed and marvelled at, but it is beyond our control. There can be no complacency with volcanic systems, and monitoring systems must continue to be provided for more volcanoes, so that Man and volcanoes can live together in the best possible way. To that extent we also live in an increasingly digital age, where data and remote information can be gathered at pace and in vast quantity. Never before has there been a better time to get into looking at volcanoes and their dynamic systems; and who knows – you may be drawn even further into the world of 'Hot Rocks'.

Glossary

Dr Volcano's A to Z of Volcanoes (including glossary of key terms used in the book)

A

'a'a (lava) [47]: a type of lava flow, from the Hawaiian for 'stony rough lava', its top surface being made of a rubbly mixture of broken pieces of lava, some with very sharp spines, often termed clinker.

accretionary lapilli [67]: balls of sequential coats of ash sticking to small objects and growing in concentric rings, caused during eruptions that interact with water.

acidic [10]: compositional term for igneous rocks and magmas that are high in silica.

active volcano [101]: a volcano that is currently erupting and/or showing significant degassing, fumarolic and possibly seismic activity. However, some definitions list as active volcanoes that have erupted in the last 10 000 years.

aeolian [62]: relating to or arising from the action of the wind, a term coined from the Greek god of the wind 'Aeolus'. In earth science it can be used to define the action of wind such as Aeolian processes, and deposits e.g. aeolian sediments.

Afar Rift [22]: part of the African rift system in the north-eastern area of Ethiopia.

agate [51]: a fine-grained microcrystalline variety of quartz, which often fills large vesicles in lava flows, producing pretty collectable stones when sectioned and polished.

air-fall deposits [???]: pyroclastic deposits formed by the settling of volcanic material (ash, lapilli etc.) through the air by gravity.

albedo [85]: the reflecting power of a surface as defined by how it reflects radiation (e.g. solar radiation), dark surfaces adsorbing radiation and light-coloured surfaces reflecting more.

Alfred Wegener [19]: (1880–1930) a German scientist, who first proposed the theory of continental drift in 1912, which hypothesised that the continents were slowly drifting around the Earth.

amethyst [51]: a semi-precious form of quartz that contains impurities which result in a beautiful purple coloration.

amino acid [85]: a type of molecule that is critical to life. Key elements of an amino acid are carbon, hydrogen, oxygen, and nitrogen.

amphibole [14]: an iron–magnesium (Fe–Mg) rich mineral found mainly in intermediate igneous rocks (*see* Table 1.1 and Figure 2.3).

amygdales [51]: vesicles filled with a secondary mineralisation, often calcite, quartz or zeolite minerals.

andesite [10]: a greyish volcanic rock, intermediate in composition between basalt and rhyolite.

anorthoclase [55]: ((Na, K)AlSi$_3$O$_8$) is part of the alkali feldspar series, in which the sodium-aluminium silicate member exists in larger proportion than the potassium-aluminium silicate.

anoxic event [85]: an event in time where the Earth's oceans become completely depleted of oxygen (O$_2$).

anthropogenic [88]: as caused by man (man-made).

Archaean [56]: the time on Earth before 2.5 billion years ago (Ga), or 2 500 million years (Ma).

Arthur Holmes [19]: (1890–1965) British geoscientist whose work included the first uranium–lead radiometric dating of rocks, contributions to our understanding of the age of the Earth, and was a great champion of the theory of continental drift long before plate tectonics were understood.

ash [59]: the smallest component of explosive volcanoes, particles less than 2 mm in diameter, composed of broken rocks and volcanic glass.

ash-fall deposits [65]: fine-grained deposits that have formed from the fallout of ash grade material through the air during an eruption.

a-type signals [97]: *see* **VT**

B

basalt [10]: A dark pasty-grey volcanic rock, with only about 40–52 per cent of silica, but relatively rich in iron, calcium and magnesium. When molten it can exceed 1000 °C. It is by far the most common volcanic rock, forming the bulk of the ocean floors, and on land it occurs in many lava flows, cinder cones, shield volcanoes and volcanic plateaux.

batholith [77]: a term (from the Greek for bathos – depth, and lithos – rock) for very large emplacements of predominantly silicic/felsic rocks such as granite in the crust. Their surface expression, where exposed, can be greater than 100 km² but often several thousand square kilometres, representing vast volumes of magma emplacement.

Barðarbunga [41]: a stratovolcano located under the Vatnajökull glacier in Iceland.

black smokers [54]: hot hydrothermal submarine vents that expel minerals rich in zinc, copper and lead as metal sulphides (in a black cloud). In some instances fossil examples of these can be found, which contain economic mineral deposits, termed massive sulphide deposits.

block and ash flow [63]: deposit from a pyroclastic flow that lacks pumice fragments, often formed from the collapse of a dome.

bomb (volcanic) [33]: large ejected magmatic material from an explosive eruption. Volcanic bombs differ from volcanic blocks in that their shape records fluidal surfaces (bombs were juvenile liquid magma when erupted).

Bowen's reaction series [13]: the set of minerals found at different compositions and temperatures for igneous rocks. As derived by *Norman L. Bowen* during melting experiments on real rocks (*see* Figure 2.3).

breakout (lava) [46]: when a new extrusion of lava breaks through the crust of an active lava flow. The lava flows advance by a series of inflated crusts and new breakouts at the front of the advancing flow.

breccias [27]: broken up coarse-grained rock made of angular clasts.

b-type signals [98]: *see* **LP**

C

Cascades [40]: mountain range that runs along the western coast of the United States of America, containing a number of active volcanoes such as Mt St Helens.

calcite [51]: mineral form of calcium carbonate (chemical formula $CaCO_3$)

caldera (collapse) [33]: the collapse structure that occurs when a volcano has evacuated a large volume of magma during an eruption. The ground above the void left by the removal of the magma collapses into it, leaving a negative valley-like structure at the surface. The term is derived from the Spanish word for cauldron. The Portuguese term 'caldeira' is sometimes used.

carbon dating [40]: a method to chronologically date material rich in carbon using the decay system of different carbon isotopes through time.

carbon gases [88]: gases rich in carbon such as carbon dioxide (CO_2), and carbon monoxide (CO).

carbonatite [56]: a type of volcanic or intrusive igneous rock that contains in excess of 50% carbonate minerals.

chert [54]: a hard sedimentary rock composed of SiO_2 in the form of microcrystalline quartz. Cherts can be formed through biological origins but also occur inorganically as a secondary chemical precipitate in rocks (e.g. nodules in chalk). Radiolarian cherts, for example, are often associated with pillow lavas, formed by the accumulation of the microscopic silica shells of radiolaria (a single-celled planktonic creature).

cinder cone/scoria cone [32]: a steep conical hill, usually less than 250 m high, with straight slopes that are initially at the angle of rest of the loose materials composing the cone. Formed above a vent when moderate explosions accumulate layers of scoria, lapilli and ash, often after sustained strombolian or fountaining eruptions.

clast (clastic rock) [38]: a fragment of a pre-existing rock or material. Clastic rocks are composed of fragments, or clasts, of pre-existing rock.

cleavage (mineral/rock) [5]: the tendency of a mineral or rock to break along specific lines of weakness caused either by the internal structure of the mineral, or by the alignment of platy minerals in the rock.

Colima (stratovolcano – Mexico) [33]: this classic volcanic cone is arguably one of the most active of the Mexican volcanoes, with a history of notable eruptions and collapses that have scarred the landscape and have even been mapped out as having flowed some 120 km from the volcano to the Pacific coast. Presently a large dome has formed in the crater.

colonnade [50]: vertical set of columnar cooling joints that develop at the base of lava flows, particularly well developed in lavas cooled by water. Colonnade occurs below entablature columnar joints in the same flow.

Columbia River basalts [27]: a large igneous province comprising continental flood basalts that erupted mainly between 17 and 14 million years ago across parts of the US states of Washington, Oregon, and Idaho.

columnar jointing [49]: concentric joints that form as commonly 5, 6, or 7-sided columns in response to the cooling and contraction of a magma, found in lava flows and sills and dykes. They are particularly well developed in lavas that are significantly aided in their cooling by water.

composite volcano [33]: *see* **stratovolcano**

conservative margin/plate boundary [24]: *see* **strike-slip**

constructive plate boundary [21]: a boundary between two of the Earth's plates where the direction of movement of those plates is away from each other, resulting in the creation of new crust.

constructive surfaces (shield volcano) [31]: low

angle layers of mainly lava flows that build up the flanks of shield volcanoes.

continental crust [12]: the layer or rocks (igneous, metamorphic and sedimentary) that make up the continents on Earth. Continental crust is relatively light and contains high amounts of silica (Si) and aluminium (Al).

continental drift theory [19]: a theory first proposed by Alfred Wegener in 1912, which hypothesised that the continents were slowly drifting around the Earth, later confirmed by plate-tectonic theory.

convergent plate margin [22]: a boundary between two of the Earth's plates where the motion of the plates is towards each other, resulting in a thickening and/or destruction of crust.

Cotopaxi (stratovolcano – Ecuador) [42]: this is the highest active volcano on our planet; this volcano's summit is laden with snow and ice, making it a potential hazard in terms of lahars and debris flows.

core (Earth) [11]: the high-density innermost part of the Earth, high in iron and nickel, comprising a liquid outer core and a solid inner core.

COSPEC or correlation gas spectrometry [98]: a UV spectrometer instrument (similar to FLYSPEC) used to detect concentrations of sulphur dioxide gases.

craton [56]: the oldest portions of continental crust.

crust [12]: the solid outer layers of the Earth forming both the continents and the ocean floors.

crustal thickening [22]: the process by which crust on the Earth's surface is thickened, for example at subduction and collision zones where continental crust is thickened through compressive forces.

cryovolcanism [57]: a cold volcano that erupts volatiles such as water (ice), ammonia, methane and sulphur, found on some of the moons in our solar system.

cryptodome [32]: swelling of the ground causing a dome-shaped structure created by accumulation of viscous magma at a shallow depth.

crystal systems [8]: the seven systems (cubic, tetragonal, orthorhombic, monoclinic, triclinic, trigonal, and hexagonal) defined in terms of internal structure and symmetry. Every mineral on the Earth fits into one of these systems.

D

dacite [10]: a pale volcanic rock, rich in silica (63–68 per cent), which is emitted at temperatures usually about 800 °C to 900 °C and is viscous and slow-moving. A common constituent of domes, it is often also involved in violently explosive eruptions.

Dallol, Ethiopia [56]: the lowest sub-areal volcano in the world at some 100 m below sea level, this volcano erupts mixtures of lava and salts. It is also

riddled with harsh acid pools, due to its interaction with salty groundwater. It has the record for being the hottest place on earth.

debris avalanche [113]: *see* **sector collapse**

Deccan Traps [85]: a large igneous province (LIP) comprising continental flood basalts that erupted most of its volume around 65 million years ago, preserved now across large parts of onshore India and offshore India and Pakistan. The province is linked to the extinction event that killed off the dinosaurs.

decompression (melting) [16]: melting of the mantle due to pressure release that is faster than the time it takes for the temperature to equilibrate, resulting in a lowering of the solidus (melting temperature).

destructive plate boundary [22]: *see* **convergent plate boundary**

diamond [59]: the hardest mineral on Earth, made entirely of carbon, with a hardness of 10 on the Mohs hardness scale. Diamonds are formed deep in the Earth at high pressures and temperatures.

diaper [77]: a balloon-like upwelling of material due to density differences where a lower, less dense material flows upwards into the overlying material, much in the same way as the wax in a lava lamp.

differentiation [12]: the changing of a liquid magma composition due to the removal of constituent elements by crystallisation.

digital terrain models [99]: a 3D model of the surface (terrain) of a particular part of the planet. Such models are usually defined by a series of xyz points on a triangular grid, which match as closely as possible the elevation at points along the terrain in question.

dike (US) [70]: *see* **dyke**

DOAS or differential optical absorption spectroscopy [98]: a method to determine concentrations of gases (e.g. sulphur and halogens at volcanoes) by measuring their specific narrow band absorption structures in the UV and visible spectrum.

dolerite [10]: classification term for a basic intrusive igneous rock that is medium-grained.

dome [10]: a rounded convex-sided mass of volcanic rock, which is usually silicic and too viscous to flow far from the vent. Often formed on stratovolcanoes towards the end of an eruption. Frequently composed of dacite, phonolite, trachyte or rhyolite.

dormant volcano [102]: a volcano that can be considered as part of an active system, e.g. one that has erupted in the last 10 000 years, but one that has not erupted for some time, and shows no clear signs of activity such as degassing and seismic activity.

ductile deformation [19]: a type of deformation where rocks deform in a plastic fashion like a viscous liquid, forming folds instead of brittle fractures and faults.

dyke (US spelling – dike) [3]: a vertical/sub-vertical planar sheet of frozen magma that has intruded pre-existing rocks, commonly cutting across the bedding in sedimentary rocks, for example. In reality dykes are rarely planar structures and can be intricately associated with sills and sill complexes as part of the plumbing system of volcanoes.

dyke swarm [73]: a symmetrical or radial set of closely spaced vertical to sub-vertical sheet intrusions, sometimes, but not exclusively, associated with igneous centres.

E

Earth's magnetic field [11]: also known as the geomagnetic field, is caused by motions in the liquid core parts of the Earth. It can be closely approximated by the field of a magnetic dipole positioned near the centre of the Earth.

effusive eruption [35]: a volcanic eruption that is not explosive but involves lava flowing out from fissures or eruptive centres.

eruption [1]: the way in which gases, liquids and solids are expelled onto the Earth's surface by volcanic action, ranging from violently explosive outbursts to effusive or hydrothermal outflow.

ejecta [36]: material/fragments that have been erupted from a volcanic vent, have travelled through the air (or possibly water) and have fallen back to the ground.

electrical resistivity [98]: the measure of how strongly a material opposes the flow of electric current through it.

EM or electromagnetic fields [98]: the physical field produced by electrically charged objects, combining an electric and magnetic field, and is one of the fundamental forces of nature. The movement of shallow magma, for example, can change the electromagnetic field at the surface.

end-Permian extinction [85]: a mass extinction of life on the planet that occurred at the end-Permian geological time boundary at around 250 million years ago.

entablature [50]: an irregular, often radiating, form of columnar jointing that occurs at the top parts of lava flows and is particularly well developed in flows that have been aided in their cooling by water. Entablature occurs above colonnade columnar joints within the same flow.

epiclastic deposits [69]: sedimentary rocks composed of re-worked volcanic material (containing less than 25% of primary volcanic material).

ERBS [???]: *see* **SAGE II**

Erta Ale, Ethiopia [54]: arguably the oldest active lava lake on the planet, Erta Ale sits in the Afar desert (also known as the Danakil depression) in Ethiopia. It is known by the local Afar people as 'The Gateway to Hell'.

eruption column [37]: the column of hot ash and volcanic gases that rise up from an explosive volcanic eruption above the vent, and can extend many kilometres into the atmosphere.

eutaxitic texture (structure) [67]: elongate structures that develop during the welding and ductile deformation within pyroclastic rocks that are associated with particularly hot ignimbrites. Commonly these are picked out by the flattening of ductile pumice clasts, forming a eutaxitic texture known as *Fiamme*.

exothermic (chemical reaction) [12]: a reaction changing one chemical or element to another, which involves release of heat as part of the reaction. The decay of one radioactive element to another, for example, is often an exothermic reaction.

extinct volcano [103]: a volcano that is never expected to erupt again. It has not erupted in historic times and there is no measurable seismic, fumarolic or other volcanic activity.

extinction events [42]: *see* **mass extinction events**

Eyjafjallajökull [3]: an ice cap (the name is Icelandic for 'island-mountain glacier') covering the volcano that erupted in early 2010, causing the 'Iceland volcanic crisis', which resulted in travel chaos around the world. It is situated in the southern part of Iceland next to the volcano Katla.

F

fallout deposits [64]: deposits formed from the fall of ash and light pumice clouds, which blanket the topography.

feldspar [14]: a group of rock-forming silicate minerals that make up as much as 60% of the Earth's crust, e.g. plagioclase and K-feldspar (*see* Table 1.1 and Figure 2.3).

felsic [48]: a term often used to describe rocks formed from magmas high in silica content, as they contain a large abundance of feldspar minerals.

Ferdinandea [33]: a submerged volcanic island, 30 kilometres (19 miles) south of Sicily, which formed a new island during an eruption in 1831, causing a four-way dispute over its sovereignty between England, Spain, France and Italy, only to disappear under the waves again ~6 months later.

fiamme [67]: a flame-like eutaxitic texture formed by the flattening of ductile pumice clasts when the central parts of hot ignimbrites deform and weld together under their own weight. These form flame-like streaks in the rock, the term *Fiamme* coming from the Italian for flame.

fire fountain [32]: a natural fountain of molten magma that is jetted into the sky through pressure caused by escaping gases as the magma reaches the surface.

fissure [21]: a crack, fault or cluster of joints, cutting deep into the Earth's crust, which may allow magma to reach the surface. A fissure usually gives rise to effusive emissions, which may be accompanied by rather more explosive eruptions forming cones of cinders or spatter known as 'fissure vents'.

fissure vent [31]: an eruptive centre located along a fissure/fault in the Earth's surface.

FLIR or forward-looking infrared radiometer [101]: *see* **infrared radiometer**

flood basalts/flood basalt provinces [27]: large-volume eruptive events, predominantly of basaltic composition, that have occurred at punctuated periods in Earth's history.

FLYSPEC or miniaturised UV spectrometry [98]: a mini-UV spectrometer instrument (similar to COSPEC) used to detect concentrations of sulphur dioxide gases.

fracture (mineral) [5]: irregular surfaces of failure in a mineral that do not follow systematic planes of failure due to the internal structure of the mineral (such as cleavage). In some instances the type of fracture may have a typical style, such as the concoidal (ring-like) fracture pattern when quartz or natural glass is broken.

fracture (rock) [48]: a discontinuity surface in a rock, such as a joint or a fault that separates the rock into two or more pieces.

fragments [17]: ash, bombs, cinders, lapilli or pumice shattered by explosions during an eruption. They are the main constituents of cinder cones and many stratovolcanoes. Also called pyroclasts and tephra.

FTIR or transform inferred spectroscopy [98]: an instrument used to measure the concentration of gases (e.g. SO_2, CO_2) using an infrared light source.

fumaroles [98]: open hole or fracture in the ground that emits volcanic gases and steam, found located around active volcanic systems.

G

gabbro [10]: classification term for a coarse-grained basic igneous rock usually found in moderate to larger intrusions.

gas escape structure [48]: internal structures found in ignimbrites that indicate the escape (upwards) of trapped and exsolving gases through the deposit.

geode [51]: a partially filled large cavity found in some basalt units, thought to represent mega vesicles. Commonly, geodes contain beautiful quartz and amethyst crystal growths or patterned agates, and are collected around the world as geo-ornaments.

geological time [42]: the whole history of the Earth, extending back about 4.6 thousand million years.

geothermal energy [107]: natural underground energy formed from the presence of a high geotherm caused by hot rocks at depth. Often manifested in thermal springs, and in some cases can be used to harness electricity through geothermal power stations.

geothermal gradient(geotherm) [15]: the rate of rise in temperature with depth in the Earth, usually measured as degrees centigrade per kilometre.

geyser [107]: a natural spring characterised by intermittent discharge of water/steam ejected as a turbulent fountain, caused by the boiling of groundwater by hot rocks. Famous examples include Old Faithful in Yellowstone, USA, and Geyser in Iceland, from which it gets its name.

glass, natural [5]: natural silica rock glass (e.g. obsidian), formed by the rapid cooling/quenching of silicate magma. Occurs commonly in lava flow settings and explosive eruptions, but can also be found in chilled margins of some shallow intrusions.

Gondwana/Gondwanaland [42]: giant supercontinent that existed on Earth around 200–280 million years ago; its breakup into separate continents was marked by some key volcanic events, making the large igneous provinces of the Karoo (~180 Ma) and the Paraná–Etendeka (~132 Ma).

GPS or global positioning satellite systems [99]: an instrument used to exactly locate a position on the ground using data collected from orbiting satellites.

GPS stations [117]: temporary or permanent postings of global positioning systems (GPS) on the ground.

grain size [10]: the size of grains or fragments that make up a rock or unconsolidated deposit.

granite [10]: classification term for a coarse-grained acidic (felsic) igneous rock rich in quartz and feldspar.

greenhouse gas [88]: a gas, such as CO_2, whose increased abundance in the atmosphere can cause a **global warming.**

H

habit [8]: a term used for the distinctive shape of a crystal form, such as tabular, prismatic, acicular, etc.

hardness (Mohs hardness scale) [8]: the scale from 1 to 10 of the relative hardness of minerals, based on what scratches what, with diamond being the hardest at 10.

harmonic vibrations (tremor) [98]: low-frequency, long-lived vibrations, with a very rhythmic signal, which often precede eruption and can last from minutes to days.

Hawaiian eruptions [35]: Basaltic eruptions like that of the Kīlauea 2018 eruption, that form lava flows that are commonly fed from fissures involving spectacular lava fountains.

hazard map [105]: a ground map compiled to highlight the areas of different hazard risk around a volcano. Hazards could include debris or lahar (mudflow), pyroclastic flow, etc.

historical time [31]: the timespan during which events have been recorded, in however fragmentary a fashion, by observers. In the Mediterranean area it may reach back 3000 years, whereas in the New World it can be less than 200 years.

Holocene Period [102]: geological period from 10 000 years ago to the present, sometimes used to define an active/dormant volcano.

Holuhraun [41]: The largest eruption on Iceland since Laki 1783, the 2014–2015 Holuhraun fissure eruption occurred just north of the Vatnajökull glacier between 31st August 2014 and 27th February 2015

hotspot (plume) [25]: term used for the area above an anomalously hot area of the mantle, manifested by an increased amount of melting and volcanoes. A hotspot generates chains of volcanoes when the plates move over it (*see* also **plume**).

hyaloclastite [54]: a volcaniclastic rock made up of fragments of rock from a millimetre to a few centimetres containing much fresh, glassy material, formed by the interaction of lava with large water bodies, e.g. lakes and the sea, or ice.

hydrothermal [89]: a term used to describe processes involving heated groundwater, usually due to hot igneous rocks near the surface (from Greek – 'hydros' meaning water and 'thermos' meaning heat).

hydrothermal vent/eruption [83]: a term used to describe eruptions of gases, steam, and hot water, without magma.

hydrovolcanic [36]: volcanism where the eruptions involve magma interacting with water or ice.

hydrovolcanic eruption [36]: a term used to describe violently explosive eruptions in which both water or ice and magma play a significant role. Such eruptions can be termed Surtseyan in shallow sea water and lakes, and hydrovolcanic or hydromagmatic on land (sometimes termed phreatomagmatic eruptions).

I

Icelandic eruption [89]: eruptions that are usually basaltic in character, which take place notably along fissures caused basically by plate divergence in Iceland. They form long rows of relatively small cones but often produce vast lava flows. Sometimes they also develop into large volcanic systems.

Iceland plume [27]: the anomalously hot mantle upwelling, hot spot, that is situated under Iceland.

igneous [9]: term (derived from Latin – igneus, meaning of fiery) used to describe rocks and processes involving naturally molten Earth materials.

igneous intrusion [70]: a rock formed by the cooling and crystallisation of molten rock (magma) that occurs underground without breaching the surface, where the molten magma has intruded another pre-existing rock.

ignimbrite [64]: the deposit from a pyroclastic density current (pyroclastic flow). They can be voluminous deposits, often covering more than 1 km³, of pumice, broken crystals, and elongated pieces of glass ('*fiamme*') in a matrix of ash. They can be welded when they are deposited at high temperatures.

infrared radiometer [101]: a thermal camera used to measure variations in heat in the infrared light spectrum.

inner core [11]: *see* **core**, **Earth**

INSAR or satellite interferometry [100]: a satellite-based radar technique that compares surface topography maps from the same area taken at different times to record any relative changes due to ground shifts (e.g. from earthquakes or shallow igneous intrusions, etc.).

J

Jökulhlaup [34]: large flood caused by meltwater from ice caps during the eruption of a subglacial volcano.

joints [48]: a natural break/fracture in a rock where the movement due to the opening of the fracture is greater than any lateral (sideways) movement along the fracture.

juvenile clast (magmatic material) [31]: refers to the parts of a deposit that are formed from the magma directly associated with the eruption that formed it, e.g. pumice, ash, scoria etc.

K

kimberlite [17]: a type of potassic igneous rock best known for containing diamonds, erupted from volcanoes that originate from great depths (up to hundreds of kilometres) in the Earth.

komatiites [56]: high magnesium (>18 wt % MgO) magmatic rocks that erupted predominantly in the Archaean and indicate very hot eruptive temperatures in excess of 1500 degrees centigrade.

L

laccolith [74]: an asymmetrical intrusion, where the roof parts of the body have been preferentially inflated in the middle and less on the outside, giving it a mushroom type appearance, and commonly circular in plan view.

lahar [10]: debris flows rich in volcanic particles and water, often called mudflows, which are very hazardous, as they can move very quickly and destroy almost anything in their path. Lahars can occur during eruptions, particularly where ice on a volcano is rapidly melted to provide an abundant source of water, or from heavy rains after an eruption, which mobilise loose volcanic debris.

Laki eruption [27]: the 1783 fissure eruption at Laki (Lakagígar) in Iceland, rich in toxic gases, which resulted in a large number of deaths in Iceland and around the world.

lamproite [80]: *see* **kimberlite**.

Landsat TM or SPOT [99]: satellite systems used for mapping the Earth's surface – Landsat TM, Thematic Mapper produces seven bands of image data. SPOT, Satellite Pour l'Observation de la Terre, produces high-resolution, optical imaging of various types.

lapilli [61]: pyroclastic particles that range in size from 2 mm to 64 mm in diameter.

lapilli tephra [60]: unconsolidated pyroclastic material in the lapilli size range.

Large Igneous Provinces (LIPs) [42]: large accumulations of igneous (volcanic) rocks that cover areas in excess of 100 000 square kilometres. They represent great volumes of magmatic material and have occurred at short-lived, punctuated periods in the Earth's history.

lateral blast [113]: a violent explosive eruption that is directed outwards to the side of a volcano, associated with a **sector collapse**.

lava [1]: molten rock or magma that reaches the surface and solidifies on cooling. Lava occurs as flows, domes, fragments within cones, and as pillows formed on the ocean floors.

lava dome [32]: a circular/semicircular dome of extruded lava that often builds in intermediate to acidic volcanoes where the lava is quite viscous.

lava flow inflation [45]: the mechanism by which lavas feed internally below an insulating crust, which allows them to travel great distances with little cooling.

lava lake [35]: an open, active lava surface that is constrained on all sides like a normal lake.

lava pathway [51]: term used for the underground series of pipes and tubes that feed distal lava flows.

lava tube [51]: a partially filled or empty underground tube where lava is travelling along, or had travelled along in the past.

LIDAR or laser scanning techniques – laser ranging technology [103]: instruments that can measure the exact 3D position of objects using lasers.

LIPs [42]: *see* **Large Igneous Province**

liquidus [16]: the line on a phase diagram that marks the maximum temperature of a magmatic liquid when it will first start to form crystals.

lithic fragments [58]: fragments of country rock and old volcanic rocks that are incorporated into a pyroclastic or volcaniclastic rock.

lithosphere [12]: the rigid outermost shell of the Earth comprising the crust and a varying amount of upper mantle, which acts as the solid part of the plates in plate tectonics.

lopolith [74]: a shallow intrusion that is circular in plan view, where the middle part sags downward so that the intrusion is shaped like a bowl or upside-down mushroom.

Lord Kelvin [12]: (1824–1907) the scientist famed for the unit of temperature named after him; for discovering absolute zero; and geologically for his work on the age of the Earth.

LP or B-type signals [98]: low frequency, long period (LP or B-Type), volcanic earthquakes that occur at shallow depths (less than 5 km), and are thought to involve the resonance of bubbly fluids in fluid-filled cracks/dykes associated with degassing or boiling and resonance of crack opening due to magma ascent.

lustre [5]: the appearance of the surface of a crystal: for example, glassy, metallic, greasy, etc.

M

Maar [80]: a German word used to describe an almost circular crater, often about 1 km across, formed mainly by hydrovolcanic eruptions. They may or may not be bordered by a ring or crescent of fine fragments, sloping gently outwards from a low crest overlooking the crater. The crater is usually filled with a small lake from which the German name is derived.

mafic [10]: a term for material rich in iron (Fe) and magnesium (Mg), e.g. the mafic minerals in Bowen's reaction series.

magma [3]: hot, mobile rock material, mainly formed by partial melting of the mantle, commonly at depths between 70 km and 200 km. It is composed of hot, viscous liquid material, also often containing crystals or rock fragments and small proportions of included gases. It is less dense than the materials surrounding it and is thus able to rise slowly towards the Earth's surface by buoyancy. If it overcomes the pressure and resistance of the rocks of the Earth's crust, it erupts in a fluid state, releasing its contained gases with varying degrees of explosive violence, and emits lava in flows or fragments.

magma chamber/reservoir [3]: a large zone of ill-defined fissures and cavities beneath a volcano, where rising magma halts for varying lengths of time. Reservoirs are most often a few cubic kilometres in volume and are situated usually between 2 km and 50 km in depth.

magnetic stripes/polarisation patterns [19]: stripes of ocean crust that contain a remnant palaeomagnetic signature of the polarity of the Earth's magnetic field when the rock was formed. As crust is created at a spreading centre along a constructive plate boundary, symmetrical patterns are recorded either side of the ridge when the Earth's polarity changes through time, creating polarisation patterns either side of the spreading centre.

mantle (Earth) [11]: the viscous differentiated layer between the Earth's core and crust, which is around 2900 km thick, and is so viscous as to be almost solid, and is composed mainly of silicates rich in iron and magnesium.

mantle plume [25]: a hot localised upwelling of mantle material underneath hot spots like Hawaii and Iceland.

mantle wedge [23]: the wedge-shaped part of a mantle between a subducting plate and the overlying plate in a subduction zone setting.

mares (the Moon) [27]: large, dark, basaltic plains (flat sea-like areas) on Earth's Moon, formed by ancient volcanic eruptions.

mass extinction events [85]: key events in Earth's history when large numbers of fauna/flora have become extinct.

mega-eruption [93]: the largest of the explosive eruptions, with a volcanic explosivity index (VEI) of 8 or greater.

melt inclusion [18]: small pockets of melt that get trapped within crystals as they are growing, which can be used to gain information about the magmatic conditions the crystal was growing in.

mica [14]: a soft, platy silicate rock-forming mineral found mainly in evolved silicic/felsic rocks (*see* Table 1.1 and Figure 2.3).

microgranite [10]: classification term used for a medium-grained (1–5 mm) silicic/felsic igneous rock characterised by the mineral assemblage of granite.

microgravity [98]: small changes in the gravitational field, e.g. due to movement of different density melts in the shallow crust.

microlite [47]: very small crystals that grow, usually during the final ascent and eruption stages of a magma.

Mid-Atlantic Ridge [21]: the mid-ocean rift that runs down the centre of the Atlantic Ocean.

mid-ocean ridge [78]: A ridge on the ocean floor where volcanic eruptions generate new oceanic crust and where two adjacent plates diverge, which defines the location of a constructive plate boundary.

mineralisation [107]: the formation/concentration of metals and other ore deposits, usually by hydrothermal processes.

mineralogy [4]: the study of the chemical, structural, and physical properties of minerals. Also used to describe a collection of minerals that make up a rock, e.g. granite mineralogy.

mitochondrial DNA [108]: molecules that contain the genetic instructions used in the development and functioning of all known living organisms; in this case the mitochondrial DNA converts the chemical energy from food into a form that cells can use.

MODIS or modern resolution imaging spectroradiometer [101]: a satellite imaging system, the Moderate Resolution Imaging Spectroradiometer (MODIS) flies onboard NASA's Aqua and Terra satellites as part of the NASA-centred international Earth Observing System.

monogenetic cone [33]: volcanic cones, such as that of Parícutin in Mexico, that are characterised by being formed by a single short-lived volcanic event.

Montserrat [40]: *see* **Soufrière Hills**

Mt Erebus [54]: an active volcano in Antarctica, the volcano was viewed as active from a distance in 1841 by James Ross, who named it *Erebus* after one of his ships. At the summit there is an active lava lake.

Mt Pelée [36]: an active volcano on the island of Martinique that is famous for its eruption in 1902 which killed about 30 000 people.

Mt Pinatubo [91]: an active volcano in the Philippines, which exploded violently in 1991 with a VEI-5–6 eruption, ejecting some 10 km^3 of magma in an event some ten times greater than that of Mt St Helens in 1980.

Mt St. Helens (stratovolcano – USA) [40]: probably one of the most famous volcanoes, this is also one of the most active ones in the Cascades range in the USA (*see* Figure 10.8). On 18 May 1980, the volcano collapsed, leading to a spectacular eruption, and resulted in a vast wasteland to the north of the volcano and 57 deaths. This eruption was very important volcanologically, as it was the first witnessed sector collapse type eruption and lateral blast. Many deposits around other volcanoes have now been re-interpreted as representing sector collapse deposits.

mud volcano [34]: term used to refer to formations (often cone/volcano-like) created by extruded muddy liquids and gases.

N

Norman L. Bowen [13]: (1887–1956) experimental petrologist from Canada, who conducted a number of melt experiments while at the Carnegie

Institution of Washington, USA. He published *The Evolution of the Igneous Rocks* in 1928, and provided the template of minerals that crystallise from rocks of different compositions at different temperatures, known as **Bowen's Reaction Series**.

nuée ardente [36]: 'glowing cloud' – a French term used to describe some pyroclastic flows, first used to describe the disastrous 1902 eruption of Mount Pelée.

Nyiragongo [54]: an active lava lake in the Democratic Republic of the Congo, situated near the town of Goma. On 10 January 1977 this lava lake drained in less than an hour, through a crack in its side. It resulted in some of the fastest ever recorded lava flow speeds, of up to 60 mph, which devastated everything in its path.

O

obduction [78]: the process in a subduction zone, by which crust is placed on top of other crust, commonly used to explain the preservation of oceanic crust that would normally be subducted in such settings.

obsidian [48]: volcanic glass. A dense, shiny black or brown, glassy and rare form of rhyolite, which rises and cools rapidly, and is usually too viscous to flow far from the vent. It forms domes, mounds and short, rugged lava flows.

oceanic crust [12]: covering between 60 and 70% of the Earth's surface, the oceanic crust is the uppermost 8–10 km of the oceanic lithosphere. It comprises shallow intrusions and lava flows that were emplaced during the creation of new ocean crust at constructive plate margins.

olivine [14]: an iron–magnesium (Fe–Mg) rich mineral found mainly in basic igneous rocks (*see* Table 1.1 and Figure 2.3).

Olympus Mons (Mars) [29]: one of the largest volcanoes in our planetary system. Olympus Mons is a massive shield volcano on Mars that rises some 27 km from the plains that surround it, and is roughly 600 km in diameter.

ophiolite/ophiolite complexes [78]: preserved sequences, often structurally controlled, which display all or some of the rocks that form the ocean crust.

oxidising [84]: the loss of electrons or an increase in oxidation state by a molecule, atom, or ion.

P

pahoehoe [21]: the term 'pahoehoe' is Hawaiian, meaning 'smooth, unbroken lava', and is one of the most common forms of basaltic lava flow, formed from fluidal low viscosity lava. The surface can often have a ropy and bulbous surface appearance, and thick pahoehoe flows can develop through a process of lava flow inflation.

palaeomagnetism [19]: the fossil record of the Earth's magnetic field that is set into crystallising igneous rocks at the time they cool.

Parícutin [106]: a monogenetic cinder cone volcano in Mexico that erupted from 1943 to 1952, building up the cone and feeding a lava flow that stopped short of fully engulfing the church of the local town, now a place of pilgrimage.

parasitic cone [33]: a small volcanic cone that lies on the side of a much larger volcano.

Peléan eruptions [36]: named after the 1902 explosive eruption of Mt Pelée in Martinique, these are explosive volcanic eruptions similar to Vulcanian eruptions, but are characterised by hot glowing clouds (termed *nuée ardente*, meaning glowing cloud) of pyroclastic flows.

Pele's hair [62]: volcanic glass threads/fibres and beads formed when small particles of molten material are thrown into the air or stretched out.

phenocrysts [50]: large crystals that are present in a fine-grained groundmass, usually indicating a phase of crystallisation at depth (in a magma chamber) before the final emplacement of the rock.

phreatomagmatic [31]: explosive eruptions involving the interaction of magma with water.

phreatoplinian eruption [67]: a large Plinian style, very explosive eruption, where significant additional water is involved.

pillow lava [53]: bulbous 'pillow'-like structures formed by effusive eruption of lava under water.

plagioclase [14]: a group of rock-forming silicate minerals that have a structure rich in sodium (Na) and/or calcium (Ca) (*see* Table 1.1 and Figure 2.3).

plate [19]: the rigid upper slabs of the Earth that move slowly around the convecting and cooling mantle. Their edges constantly diverge, or converge and plunge beneath each other. All are composed of oceanic crust and some also carry continental crust. Between 10 and 15 major plates, and a similar number of micro plates are generally recognised.

plate tectonics [19]: the theory, which built on that of continental drift, that splits the Earth's surface into a series of plates that can move slowly and interact at three types of plate boundary: convergent (destructive), divergent (constructive) and strike-slip (conservative) *see* Figure 3.2.

Pliny the Younger [36]: (AD 61–112) considered by many as producing the first scientific description of a large explosive Plinian eruption (which now bears his name). Pliny the Younger witnessed and documented the AD 79 eruption of Vesuvius that destroyed Herculaneum and Pompeii, and also killed his uncle Pliny the Elder.

Plinian cloud [36]: the classic mushroom-like cloud formed by a Plinian eruption. Described by Pliny the Younger as being a shape like that of an umbrella pine, a tree that can be found growing around Pompeii today.

Plinian eruption [33]: named from the famous AD 79 eruption of Mt Vesuvius, described by Pliny the Younger. These eruptions are very powerful and result in a large column and cloud of ash and ejecta that travels high up into the atmosphere (up to 55 km) and stretches out into a mushroom-like shape where it reaches the stratosphere. They are generated mainly by silicic magmas, although they can develop from hydrovolcanic eruptions of basalt. The gas and dust expelled to the stratosphere often form an acidic aerosol that can modify the weather over large tracts of the Earth.

plug [74]: a circular/semi-circular body of igneous rock formed in the feeding pipe/vent of a volcano.

plume (mantle) [25]: *see* **hotspot (plume)**

pluton [77]: a large body of igneous rock, commonly of granitic composition, which has crystallised at depth in the Earth's crust.

plutonic [10]: igneous rocks that have crystallised at depth beneath the Earth's surface.

Popocatepetl (stratovolcano – Mexico) [33]: this volcano towers over Mexico City (population 8.8 million), and is capable of pyroclastic flows, sector collapse and large volume lahar deposits. The second highest volcano in North America, it has been recorded as far back as the Aztecs as having large Plinian eruptions (e.g. AD 800). Currently the volcano is constantly degassing and is monitored for changes that could spark a new eruptive period.

'pop-up' volcano [33]: an emergent volcano that periodically rises up out of the ocean during an eruptive phase and is then eroded back by the sea. A classic example is **Ferdinandea** off southern Sicily.

primordial soup [84]: the large ocean of chemicals from where life began in early Earth's history.

pumice [61]: very pale volcanic fragments riddled with gas holes, formed by the expansion of contained gases as the magma reaches the surface and explodes very violently over vast areas during an eruption. Most pumice floats on water and sometimes forms ephemeral floating islands after eruptions at sea. It varies from small fibrous chips to knobbly lumps and often resembles solidified foam. It is commonly expelled in eruptions of rhyolite, dacite, trachyte or phonolite.

pyroclastic density current (pyroclastic flow) [10]: term for a turbulent current of hot particles, rock fragments and gas, it is commonly termed 'pyroclastic flow'. When the current has a low concentration it is sometimes referred to as a 'pyroclastic surge'. There is a complete continuum

between high- and low-concentration pyroclastic density currents, and they can mix between the two during an eruptive phase. Other old terminology includes the French word '*nuée ardente*' for fiery cloud.

pyroclastic surge [63]: a term used to describe a low-concentration turbulent current of hot particles, rock fragments and gas. A low-concentration pyroclastic density current.

pyroxene [14]: an iron–magnesium (Fe–Mg) rich mineral found in basic to intermediate igneous rocks (*see* Table 1.1 and Figure 2.3).

Q

quartz (SiO2) [14]: common rock-forming mineral found in felsic igneous rocks, made of a continuous framework of SiO_4 silicon–oxygen molecules (*see* Table 1.1 and Figure 2.3).

R

radioactive elements [12]: elements that are unstable and change to other elements by a process called radioactive decay. The atomic nucleus of an unstable atom loses energy by emitting ionising particles, a reaction that gives off heat.

radiolarian chert [54]: a silica-rich material formed by the silica tests of microfossils called radiolaria.

reducing [84]: the gain of electrons or decrease in oxidation state by a molecule, atom, or ion.

resurgent dome [32]: a volcanic dome that grows in the centre of a recently formed eruption crater/caldera as the late stages of volcanic activity.

rheomorphic ignimbrites [67]: very hot ignimbrite deposits that are so hot that they start to re-form back into a lava.

rhyolite [10]: a pale volcanic rock very rich in silica (69–75 per cent), which is usually emitted at temperatures about 700–800°C and commonly forms extensive pumice and ash flows when expelled as fragments, but it also gives rise to viscous lavas forming domes and stubby flows.

rilles (the Moon) [27]: channel or levee-like features on the Moon thought to be formed by lava flows/tubes.

Ring of fire [23]: the circle of volcanoes that rings the pacific ocean.

S

SAGE II – Stratospheric Aerosol and Gas Experiment II [92]: a NASA satellite-based tool, on the Earth Radiation Budget Satellite (ERBS), used to map the distribution of aerosol, ozone, water vapour and nitrogen dioxide.

Santorini [4]: also known as Thera, a volcanic island located in the southern Aegean Sea. Its current structure marks out a caldera that formed by a massive explosive eruption that is thought to have had an impact on the Minoan civilisation in the Mediterranean. Recent activity has produced the Kameni islands in the centre of the caldera, which last erupted in 1950.

scoria [32]: primary pyroclastic material made up of vesicular basalt or andesite, formed by explosive eruptions of more basic magmas.

seamount [25]: a volcanic mountain found below sea level, rising from the ocean floor. Active seamounts could eventually form new volcanic islands. They may also be extinct submerged remains of old volcanoes.

secondary minerals [51]: minerals formed by post-magmatic processes, e.g. hydrothermal, and by alteration.

sector collapse (debris avalanche) [113]: a landslide formed at the start of a volcanic eruption, where one side of the volcano collapses. This can be triggered by over-steepening of the volcano, and/or by earthquakes, and can cause an asymmetric eruption known as a **lateral blast**.

seismic activity [114]: earthquakes caused by shaking of the Earth, which can be caused by movement along a fault, migration of magma, and explosions during eruptions.

seismic signals [17]: the specific wave signals (shock wave, pressure wave) caused by seismic activity.

seismometer [96]: instrument used to record seismic signals from earthquakes.

shield volcano [31]: a large, gently sloping (shield-like) volcano, composed mainly of fluid basaltic lava flows with relatively few fragmented layers, emitted from clustered vents.

Siberian Traps [27]: one of the largest Large Igneous Provinces made up of continental flood basalts that cover large tracts of Siberia. This particular massive volcanic outpouring occurred ~250 million years ago and is implicated in the end-Permian extinction (~250 Ma)

silica [10]: the molecule, formed of silicon and oxygen (SiO_2), that is a fundamental component of volcanic rocks, and is the most important factor controlling the viscosity (fluidity) of magma. Other things being equal, the higher the silica content of a magma, the greater its viscosity.

silicates [12]: the most important class of rock-forming minerals, which all contain both silica (Si) and oxygen (O) within their structure.

silicic [33]: descriptive term for rocks rich in silica, such as granite.

sill [3]: tabular sheet-like intrusion which is horizontal/sub-horizontal, concordant with structures present in the country rock, e.g. sedimentary bedding. In reality, sills are rarely simple planar structures and can be found as linked sill complexes with several levels of intrusion, as saucer-shaped intrusions, and in close association with dykes.

slab melting [23]: a process of forming melt in subduction zones where fluids driven off the subducting slab enter the mantle and lower the melting temperature.

solidus [15]: the temperature at which a rock will start to melt. Often displayed as a line with depth depicting the change in the wet-solidus with depth (pressure) *see* Figure 2.6.

Soufrière Hills [40]: the Soufrière Hills (French for 'sulphur') volcano is an active stratovolcano on the island of Montserrat. It became active in 1995, after a long period of dormancy, destroying the capital city, Plymouth, and causing widespread evacuations and relocation of its inhabitants.

stalactite (volcanic) [51]: downward-growing drips of lava found on the roof of lava tubes as trails of cooled falling magma.

stalagmite (volcanic) [51]: upward-growing mounds of magma that build up as a result of drops of magma falling from the roof of a lava tube.

stratovolcano [33]: a type of volcano, found mainly at destructive plate boundaries, that rises up from shallow slopes at the base to form steep-sided tops of the volcano, resulting in a cone-shaped mountain, often with a surprisingly small crater at the top. Eruptions from such volcanoes can be violent and explosive with Plinian-type eruptions, and are constructed mainly from more silicic/felsic magma types.

streak [5]: the characteristic colour left by some minerals when they are scratched on the back of a porcelain tile.

strike-slip or conservative margin [24]: a plate-tectonic boundary where the two plates are moving side by side against each other. No crust is gained or lost at such a boundary.

Stromboli [3]: known as 'the lighthouse of the Mediterranean', this volcano is one of the Aeolian Islands off northern Sicily. It erupts regularly, every 15–20 minutes, with a characteristic eruption style, 'strombolian', to which it lends its name.

Strombolian eruption [32]: a characteristic eruption style (named after the volcano Stromboli, where the eruptions are common) consisting of small amounts of hot basic magma that are erupted explosively. Continued eruptions build up a scoria cone, and when viewed in low light/night-time, a fountain-like (similar to roman candle fireworks) trajectory of the hot, glowing scoria can be seen.

subducting slab [22]: the downward-going cold lithosphere (including ocean crust) in a subduction zone.

subduction [22]: the process by which one tectonic plate is sunk under another at a destructive plate boundary.

subduction zone [23]: a zone where two plates converge and one plunges beneath the other into the mantle. The subducted slab releases volatiles that stimulate melting in the wedge of mantle above it, which helps form volcanoes. The plunging action of the slab also generates deep-seated violent earthquakes.

subglacial volcanoes (glaciovolcano) [34]: volcanoes that are under significant ice cover, usually ice caps, many examples being those found on Iceland (e.g. Grímsvötn caldera, Katla caldera), and in Antarctica.

submarine volcanoes [33]: volcanoes that erupt under the sea; many are located along the plate boundaries, and are thought to account for some 75% of annual volcanic output.

sulphur gases [88]: key gases containing sulphur, such as sulphur dioxide (SO_2) and hydrogen sulphide (H_2S).

supercontinent [42]: a large continent made up when the configuration of continental plates on the Earth brings them together. Examples of supercontinents that have occurred in Earth's history include Gondwana and Pangaea.

supervolcano [42]: a volcano that has the capability of erupting more than a thousand cubic kilometres of magma in a single event (100–1000 times larger than historic eruptions).

Surtseyan eruption [37]: an eruption that takes place in shallow lakes or seas where water can enter the vent, mix with the rising magma, and repeatedly form steam that shatters the magma into fine fragments often called tuffs. Without such water interference, most of these eruptions would probably be Strombolian in nature. One of the major types of hydrovolcanic eruption.

Système Probatoire d'Observation de la Terre (SPOT) [99]: see **Landsat TM**.

T

tephra [60]: term (of Greek origin) used to describe pyroclastic accumulations forming unconsolidated deposits.

tiltmeter [99]: an instrument used to measure tilting of the ground due to inflation during shallow magma emplacement.

Toba [108]: this volcano, located in Sumatra, Indonesia, produced arguably the largest eruption on the planet in the last 2 million years, 'the Toba eruption', which occurred approximately 71–72 000 years ago. This was a VEI-8 eruption which involved some 2800 km³ of magma, and is thought to have caused significant impacts to the planet, including a direct effect on the evolution of Man.

TOMS or total ozone mapping spectrometer [101]: the Total Ozone Mapping Spectrometer, launched in July 1996 onboard an Earth Probe Satellite (TOMS/EP), continues NASA's long-term daily mapping of the global distribution of the Earth's atmospheric ozone.

tsunami [106]: a Japanese term used to describe huge, rapidly moving sea waves generated by violent eruptions or earthquakes. They increase in size and speed as they reach shallow water and often cause much damage and death on nearby coasts.

tuff [60]: a pyroclastic deposit (consolidated) that contains a significant component of volcanic ash (>75%).

tuff cone [37]: a steep, squat conical hill, usually less than 300 m high, composed of innumerable thin layers of fine fragments with a deep, wide crater, formed above a vent by Surtseyan eruptions.

tuff ring [67]: a low-profile apron of tephra surrounding a wide crater, commonly found where the crater is formed from an eruption involving water-saturated sediment or groundwater.

tumulus [47]: a bulge or spot-like mound on a lava flow caused by the extrusion of over-pressured lava through the crust of the flow.

tuya [34]: a flat-topped, steep-sided volcano formed when lava erupts through a thick glacier or ice cap. Named after Tuya Butte, northern British Columbia, Canada.

U

umbers [54]: natural brown earths containing iron oxide and manganese oxides, found associated with hydrothermal deposits (used as a brown pigment in paint).

V

vent [27]: the usually vertical conduit or pipe up which volcanic material travels from the magma source to the Earth's surface.

vesicle [48]: a circular cavity found in volcanic rocks (and some shallow intrusions) as the result of gases exsolving from the melt to form bubbles. Vesicles can often be deformed as a lava flow cools, and are sometimes later filled with minerals, forming **amygdales**.

vesicle stratigraphy [51]: alignment of vesicles into repeated bands (like stratigraphic units), found commonly in lava flows, but which can also occur in some shallow intrusions.

vesiculation (vesiculate) [98]: the formation of bubbles (vesicles) in a magma/lava.

vesiculation event [51]: a punctuated event, usually a loss of pressure in a volcano or lava flow, which causes the magma/lava to vesiculate.

Vesuvius [106]: on the eastern coast of the Bay of Naples, this is one of the world's most famous volcanoes. It gave rise to the first detailed description of a volcanic eruption by Pliny the Younger, when its Plinian eruption of AD 79 destroyed Herculaneum and Pompeii, and it is visited by millions of people each year as a major tourist attraction. Last erupted in 1944.

viscosity [46]: a measure of the resistance of a fluid that is being deformed. High viscosity implies a sticky fluid and low viscosity a runny fluid.

VOG [89]: a term used to describe a dense fog caused by volcanic gases, e.g. sulphur dioxide and moisture in the atmosphere. A combination of the terms *volcanic* and *smog*.

volcanic [1]: describes igneous processes that are occurring at the Earth's surface; derived from the name of the Roman god of fire – Vulcan.

volcanic ash [60]: tiny particles of juvenile magmatic glass formed during the fragmentation of a magma in an explosive volcanic eruption. Term used for pyroclastic material less than 2 mm in diameter.

volcanic explosivity index (VEI) [38]: a scale derived to provide a relative measure of the explosiveness of volcanic eruptions, as measured by a number of factors including eruption column height, volume of eruption, etc.

volcanic gas [60]: the volatile component of magma, mainly including steam, carbon dioxide, sulphur dioxide and smaller amounts of chlorine and fluorine. As the magma approaches the Earth's surface, the gases are exsolved and can become the chief factor in the violence of eruptions.

volcanic winter [108]: a cooling in the atmosphere caused by the eruption of volcanic gases and particles that reduce the sunlight reaching the surface of the planet, causing the cooling after an eruption.

volcaniclastic sediments [54]: deposits composed mainly of grains and clasts derived from volcanic activity, and can be primary and/or reworked and re-deposited.

volcano [1]: a hill or mountain formed around and above a vent by accumulations of erupted materials such as ash, pumice, cinders or lava flows. The term refers both to the vent itself and to the often cone-shape accumulation above it.

VT or A-type signals [97]: signals from volcano–tectonic (VT) earthquakes that have a sudden start and last for a short time.

Vulcano [36]: one of the Aeolian islands off northern Sicily, this is the source of the word 'volcano', as the island was thought to be the home of Vulcan, the god of fire (according to Roman legend). It last erupted in 1890 and is famed volcanologically as the first place 'vulcanian' eruptions were described, and from which they also got their name.

Vulcanian eruption [36]: first used to describe the 1888–1890 eruptions on the island of Vulcano. These eruptions form explosions like cannon fire at irregular intervals, and contain a large component of non-juvenile material. Vulcanian eruptions have been attributed to the interaction of the magma with groundwater, known as hydrovolcanic eruptions.

W

welded ignimbrite [66]: a pyroclastic deposit that has been solidified, 'welded' together due to compaction and fusing of the hot primary volcanic material. Usually associated with the compaction of pumice clasts to form **fiamme**.

wet-solidus [16]: the temperature in the presence of water at which a rock will start to melt. Often displayed as a line plotted against depth, depicting the change in the wet-solidus with depth (pressure).

white smoker [80]: plumes of hot water and white particles that erupt from hydrothermal vents. White precipitates are rich in barium, calcium and silicon, and eruption temperatures are lower than that for 'black smokers'.

X, Z

xenocryst [58]: a crystal incorporated into a magma from a pre-existing unrelated rock. The crystal is mechanically incorporated, usually due to partial melting of country rocks during magma ascent.

xenolith [17]: a rock fragment found in a magma from a pre-existing unrelated solid rock. The fragment is mechanically incorporated into the magma by the breaking and/or partial melting of country rocks during magma ascent.

zeolites [51]: a group of alumino-silicate minerals that can have a wide variation of chemistry, often found associated with secondary mineralisation in igneous rocks, particularly basalts.